QUANTUM MECHANICS AND ALGORITHMS

An Algebraic-Geometric Perspective

QUANTUM MECHANICS AND ALGORITHMS

An Algebraic-Geometric Perspective

Yair Shapira

Technion – Israel Institute of Technology, Israel

World Scientific

NEW JERSEY • LONDON • SINGAPORE • BEIJING • SHANGHAI • HONG KONG • TAIPEI • CHENNAI • TOKYO

Published by

World Scientific Publishing Co. Pte. Ltd.

5 Toh Tuck Link, Singapore 596224

USA office: 27 Warren Street, Suite 401-402, Hackensack, NJ 07601

UK office: 57 Shelton Street, Covent Garden, London WC2H 9HE

Library of Congress Cataloging-in-Publication Data
Names: Shapira, Yair, 1960– author
Title: Quantum mechanics and algorithms : an algebraic-geometric perspective /
Yair Shapira, Technion--Israel Institute of Technology, Israel.
Description: New Jersey : World Scientific, [2026] | Includes bibliographical references and index.
Identifiers: LCCN 2025054113 | ISBN 9789819824588 hardcover |
ISBN 9789819824595 ebook for institutions | ISBN 9789819824601 ebook for individuals
Subjects: LCSH: Quantum theory--Mathematics | Quantum computing | Quantum cryptography
Classification: LCC QC174.17.M35 S53 2026
LC record available at https://lccn.loc.gov/2025054113

British Library Cataloguing-in-Publication Data
A catalogue record for this book is available from the British Library.

For any available supplementary material, please visit
https://www.worldscientific.com/worldscibooks/10.1142/14626#t=suppl

Desk Editors: Eshak Nabi Akbar Ali/Muhammad Ihsan

Typeset by Stallion Press
Email: enquiries@stallionpress.com

Preface

This book introduces quantum mechanics from a new refreshing point of view: not only physical (as in Planck's approach) or analytical (as in Schrodinger's equation) or algebraic (as in Bohr's theory) but also combined: all of the above, plus geometry. This introduces quantum mechanics from scratch, right into its proper place: at the front of modern science and technology.

Mathematical physics uses two possible approaches. Classical physics is mostly analytical-geometrical. Quantum mechanics, on the other hand, is much more algebraic and abstract. How to make it more visual and accessible? This book offers a compromise: from classical physics, it borrows geometry and uses it in quantum mechanics too. This helps draw a new picture, in new polar coordinates: time–energy. This way, the dynamics gets geometrical too: circular (or angular): round and round. Moreover, the Schrodinger equation takes its geometrical face too. This is how algebra and geometry join forces to explain quantum mechanics from a new point of view, with interdisciplinary connections and applications.

So, this book introduces quantum mechanics in a new light: visual and geometrical, easy to see and grasp. This puts quantum mechanics quite firmly on top of three robust mathematical legs: linear algebra, calculus, and geometry. Thanks to the new geometrical picture, physics follows ever so smoothly and naturally from math. Even dry formulas get alive, with an apparent physical meaning.

Not only that, but quantum mechanics also gets embedded right in the heart of applied science and engineering, with applications in machine learning, image processing, and cryptography. Indeed, modern technology can borrow a few fundamental principles from physics: stability, minimum energy, and more. This is how physics and engineering go hand in hand and shed light on one another. Moreover, although no quantum computer is available as yet, there are already a few (highly inspiring) quantum algorithms that could start a complete technological revolution.

Where can we use the new algebraic-geometrical framework? For one, use it to model spin and polarization, relevant in quantum computing. Moreover, our (theoretical) quantum computer takes its own geometrical face too: a (virtual) hypercube, ready for quantum FFT (recursive or not), and Shor's factoring algorithm in cryptography.

How is quantum computing related to quantum mechanics? Through two fundamental concepts: entanglement and its entropy. Here too, geometry plays a major role. Indeed, entanglement leads to an algebraic-geometrical tool: the tensor product. This helps design the (virtual) hypercube, on which quantum computing takes place.

Moreover, thanks to the concept of entropy, we get to see the fundamental unit of data: the bit. In quantum computing, this is even extended to the qubit. Thanks to entanglement entropy, we even get to see how information flows from one random variable to another. This is the key to entanglement, and quantum computing, in their full geometrical picture.

Thanks to this new framework, this book can serve as a textbook in an (undergraduate) physics course on quantum mechanics. Moreover, thanks to the new interdisciplinary connections in it, it can also serve as an (additional) textbook in a CS course on cryptography with quantum computing.

This book contains seven parts. Part I introduces quantum mechanics from scratch and illustrates its dynamics in our new time–energy coordinates. Moreover, it highlights the principle of minimum energy, useful in machine learning and image processing too. Part II introduces entropy in physics and computing, including entanglement entropy. Part III introduces spin and polarization, useful in quantum computing too. Part IV introduces practical algorithms in cryptography. Part V introduces important quantum algorithms that have a potential to transform the entire field of cryptography. Part VI introduces Feynman diagrams algebraically geometrically, with elements from quantum field theory. Finally, Part VII implements a few practical algorithms in C++, with a complete code, explained in detail, line by line.

This is written in 19 chapters. Five chapters are written in a unique style. They teach you physics in an active way: through exercises, followed by hints and solutions. This way, you get to think for yourself and discover the theory for yourself, before reading on it explicitly.

This book requires prerequisites in math only: linear algebra and calculus. It can be used as a textbook in two undergraduate courses (in physics/CS/EE/math departments):

- Introduction to quantum mechanics and its applications.
- Introduction to cryptography with quantum computing.

Roadmap: How to Read the Book?

Well, this depends on who you are:

- If you are a physicist/student/teacher, then do the following:
  - Start from Chapters 2–4, to get introduced to quantum mechanics.
  - Then, go to Part II on entropy and entanglement.
  - Then, follow to Part III on spin and polarization.
  - This will get you ready for quantum algorithms (Part V) and quantum field theory (Part VI).
- If, on the other hand, you are a computer scientist/engineer, then do the following:
  - Start from Chapter 1 on machine learning and energy, with applications in image processing (Chapter 5).
  - Then, skip to Part IV on cryptography.
  - Then, follow to Part V on quantum algorithms in cryptography.
  - Finally, conclude with the detailed C++ implementation (Part VII).

Enjoy your reading!

Yair Shapira
Haifa, Israel

Contents

Part III Angular Momentum and Spin

Part IV Algorithms in Cryptography

Part V Quantum Algorithms in Cryptography

Part VI Feynman's Theory: An Algebraic Point of View

Part I: Introduction to Quantum Mechanics

What is AI (artificial intelligence)? AI is a mathematical approach to process data automatically. In practice, it processes big data, in an optimal way. In a sense, we "teach" our computer how to solve problems for us, for not only the present data but also any other data to come. For this, we develop algorithms, useful for not only a specific project but also many others to come. In this sense, we actually "train" our computer to encounter any future data, and solve the problem efficiently.

In this part, we see how AI is related to physics. Thanks to the laws of nature, we'll learn how to increase stability in AI, and minimize energy, as in many physical processes, including the flow of heat (modeled in the heat equation). Better yet, AI can benefit from not only classical but also quantum mechanics.

How is AI different from the human mind? Well, man speaks a natural language: not only logical and rational but also intuitive and emotional. This kind of language developed quite spontaneously, to help man live his/her daily life, and take care of his/her basic needs. For this, man must communicate and form a community, society, and indeed a nation. This is why a human language must be based on empathy between people.

Still, our brain is not big enough to store (and process) big data, often required in modern technology. For this, we need AI. In machine learning, for example, we use our data to model a random process. This is quite practical: it will help take new instances (not in our data) and predict their future behavior in advance. In his classical book *I Robot*, Isaac Asimov already predicted AI, along with its problems and paradoxes.

To do AI, what can we learn from physics? A lot! Especially from quantum mechanics. In fact, both AI and quantum mechanics use the same principle, well known in physics and applied math: stability and minimum energy. After all, human energy is particularly dear to us: we want to have fun, not spend too much time on a tedious technical work.

To see this, quantum mechanics is ideal. In it, energy is discrete, not continuous: it comes in packets, or quanta. This way, energy can't take every value but only a sequence of discrete values: energy levels.

How to visualize this geometrically? This is the main element in this book. Thanks to math, we have a great tool: the complex plane. In physics, this models the phase plane: the position–momentum plane. In the quantum microscale, however, this is not quite physical but more mathematical. After all, due to the uncertainty principle, position–momentum can't be measured simultaneously. Still, with some uncertainty, they can.

So, for theoretical purposes, our new phase plane is quite useful. In it, we can now define a new complex function: a position–momentum wave function. Once squared (and normalized), it makes a new distribution (probability density), ready for integration (in 1-D, and even 2-D). In this sense, it has a lot of theoretical value: it encapsulates an *a priori* information about the (original) physical state (with no measurement or experiment that could change or harm the original state).

In this plane, how do energy levels look like? Well, they make distinct circles of bigger and bigger radius. This is quantum mechanics for you: only discrete circles are allowed, not the area in between. This is how geometry mirrors physics: it gives us new polar coordinates: energy is radial, and time is angular (circular). This will lead to dynamics and to the Schrodinger equation. This also mirrors our good old Cartesian coordinates: position–momentum are related by the Fourier transform and time–energy by the Fourier series. Moreover, time–energy obey their own uncertainty principle: just like you can't know position and momentum simultaneously, you also can't know both time and energy simultaneously.

High energy is rare. This gives us a practical guideline, useful in AI too: better drop high energy altogether and focus on low energy only. In AI, this means to filter high variation out and use a smooth function only. But now, we know even better: from our big data, we can extract a lot of valuable information about smooth modes and design them well. In image processing, for example, we can use not only low Fourier modes but also local cosines and wavelets. Moreover, in classification algorithms, this principle leaves only a few (active) degrees of freedom, with good stability, and no noise or overfitting. After all, this is all we can hope to achieve: learn about the original (random) process (and its true features) and model it well in practice, to help predict future instances too. Even with noisy data (full of errors), our model will still be stable enough, and our predictions will still be within reason.

Chapter 1

Background: Machine Learning and Energy

In this chapter, we give some background in machine learning, from a new point of view: PDEs (partial differential equations) in computational physics. In fact, from the heat equation, we'll obtain radial basis functions, suitable to expand our unknown function. This is not only efficient but also stable. After all, in nature, a physical process must be stable and approach a steady state: an equilibrium, with minimum energy.

This is not limited to physics but is relevant in data science as well. To see this, we'll borrow an important principle from physics: drop (unstable) high energy and focus on low energy only. In machine learning, this will take a practical face: reducing bias, error, and overfitting.

This is how mathematical physics can contribute to modern technology, and vice versa. Later on, we'll study energy levels (and their minimum) in quantum mechanics too. This is how physics guides us to develop practical algorithms, with a great benefit to industry, finances, data, and the entire economical environment.

1.1 Machine Learning Algorithms

1.1.1 *How to Model a Random Process?*

In machine learning, we often have N pieces of data (each with its own value). From these, we wish to learn about the underlying (random) process and its true nature. This will help us take new points (*not* in our data) and predict their value in advance. Later on, we'll see a few interesting applications in image processing. We'll also look at two simple examples: throwing a coin or a dice (Figures 6.1 and 6.2).

How will the process behave on new instances? What values will it assign to them? How could we possibly tell? After all, these new points are not yet in our data! Are we expected to be prophets? No, we are only expected to design a stable model that will uncover the major features that characterize the process. These are the (unknown) parameters (or degrees of freedom). Once uncovered, they will be highly likely to model the original process well and help predict the future. No need to be a prophet but just to have a stable algorithm, with no overfitting [1].

1.1.2 *Harmonic Extension*

Define the dimension: d (a natural number). This way, each point $x \in \mathbb{R}^d$ will have a few coordinates:

$$x \equiv \left(x^{(1)},\ x^{(2)}, \ldots, x^{(d)}\right) \in \mathbb{R}^d.$$

This will be our linear space. What will our data be? First, our data will contain N distinct points in $\mathbb{R}^d$:

$$x_1,\ x_2,\ x_3, \ldots, x_N \in \mathbb{R}^d.$$

This way, for $1 \le n \le N$, each point x_n will have d coordinates. Besides, our data will contain more: each point x_n is also assigned a real value:

$$b_n \in \mathbb{R}.$$

Together, these b_n's make the right-hand-side vector:

$$b \equiv (b_1,\ b_2,\ b_3, \ldots, b_N)^t \in \mathbb{R}^N.$$

All these are our data. Our job is to extend the data smoothly: design a smooth function

$$f : \mathbb{R}^d \to \mathbb{R}$$

that agrees with the data:

$$f(x_n) = b_n, \quad 1 \le n \le N.$$

How to do this? In Chapter 13 in Ref. [25], it is suggested to use the x_n's as nodes in a high-order (d-dimensional) finite-element mesh and solve the Laplace equation in it (with homogeneous Neumann boundary conditions and with the b_n's as inner Dirichlet conditions). This gives a big linear system of the form

$$Af = \mathbf{0},$$

subject to

$$f(x_n) = b_n, \quad 1 \le n \le N.$$

Here, $\mathbf{0}$ is a (quite long) zero vector, A discretizes the Laplacian:

$$A \doteq -\triangle,$$

and f is the unknown function: a smooth piecewise-polynomial function on the mesh. Unfortunately, even for a moderate d, this is too big to assemble and solve numerically. Better use a more analytic approach: a special kind of basis functions that solve the heat equation (not only numerically but also analytically).

1.2 Green Function: Kernel

1.2.1 *The Heat Equation*

The heat equation is a parabolic PDE:

$$u_t = \triangle u,$$

where u depends on both time t and space $x \in \mathbb{R}^d$. To solve for u, use an analytic tool: the Green function. This is a kernel: it depends on not only time but also two points in space:

$$x, \ \tilde{x} \in \mathbb{R}^d.$$

These points have a different status: x is a variable and $\tilde{x}$ is a fixed parameter.

1.2.2 *The Gaussian*

In our case, in particular, the Green function is the Gaussian:

$$G(t, x, \tilde{x}) \equiv t^{-d/2} \exp\left(-\frac{\|x - \tilde{x}\|^2}{4t}\right).$$

This indeed solves the heat equation in $\mathbb{R}^d$, with a special kind of initial data: at the initial time $t = 0$, there is a pulse of heat: a delta function, concentrated at one particular point:

$$\tilde{x} \in \mathbb{R}^d.$$

This kernel will tell us how the initial pulse spreads out over time. But this is not limited to one pulse only: you can now superpose (sum) a few pulses, and even integrate over $\tilde{x}$ (as in a convolution), and solve for any initial distribution as well.

1.2.3 *Radial Basis Functions*

How to use this in machine learning? Look at an interesting time: $t = 1$. (This is an exponentiation. This is how an initial pulse develops into a new heat wave.) Now, at $t = 1$, look at the Gaussian kernel:

$$g(x, \tilde{x}) \equiv G(1, \sqrt{2}x, \sqrt{2}\tilde{x}) = \exp\left(-\frac{\|x - \tilde{x}\|^2}{2}\right)$$

(Figure 9.1). This is a radial basis function, centered at $\tilde{x}$. How to pick $\tilde{x}$? Well, to solve our original problem, pick $\tilde{x}$ as one of our original points:

$$\tilde{x} \leftarrow x_m \quad 1 \le m \le N.$$

This will help expand our (unknown) function f as a new linear combination of radial basis functions (centered at our original points):

$$f(x) \equiv \sum_{m=1}^{N} w_m g(x, x_m),$$

where

$$w \equiv (w_1,\ w_2,\ w_3, \ldots, w_N)^t \in \mathbb{R}^N$$

is the vector of (unknown) coefficients. How to solve for w? Forget about the above A. Instead, solve (iteratively) the new linear system

$$A^{-1} w = b,$$

where A^{-1} is a new matrix, whose (n, m)th element is already available:

$$\left(A^{-1}\right)_{n,m} \equiv g(x_n, x_m) \quad (1 \le n, m \le N).$$

Thanks to this new A^{-1}, we can now deduce: what is the new A used here? Not quite the discrete Laplacian any more. Indeed, A^{-1} discretizes now our Green function: the integral operator that solves the heat equation. Thus, our new A must discretize now the heat differential operator:

$$A \doteq \frac{\partial}{\partial t} - \triangle.$$

This way, the inverse mapping

$$A^{-1} : w \to b$$

will indeed solve the heat equation for us: it will give us b (the solution at $t = 1$) in terms of w (the initial pulses at $t = 0$). Is this stable? Well, we'll soon see that A^{-1} has a good algebraic property: SPD. Thus, our new A must be SPD as well. This is a good sign: our new A stands for a stable numerical scheme, modeling a real physical process: a heat wave, spreading out, with increasing entropy, as in the second law of thermodynamics.

1.3 Principle of Minimum Energy

1.3.1 *Minimum Energy*

But here we are not quite interested in the heat equation or any initial pulse that dies out over time but more in (inner) Dirichlet constraints that nail f at N given points (and force f to have a certain value there), not only initially but also forever! On second thought, why be so pedant? After all, our data are often noisy and inaccurate. Why force them too strictly? Better drop some (unimportant) constraints and minimize energy subject to a few important constraints only! How to do this?

Let's start from the beginning. What is w? Well, w helps expand f in terms of radial basis functions (and b in terms of columns of A^{-1}). For this, w contains N coefficients, for N vectors: the columns of A^{-1}. But now, our main priority is not to expand b exactly any more but to minimize energy.

How to write the energy? In general, this is a quadratic form: $(A\cdot, \cdot)$. What vector shall be plugged in? Should it be w itself?

Not quite. After all, w is still in the wrong basis: columns of A^{-1}. We must first transform w back to the standard basis (columns of the identity matrix). For this, look at w again and apply A^{-1} to it. Thus, in our quadratic form $(A\cdot, \cdot)$, we must plug $A^{-1}w$ (rather than w itself). So, we actually need to minimize

$$(AA^{-1}w, A^{-1}w) = (w, A^{-1}w),$$

subject to

$$A^{-1}w = b.$$

These are our (inner) Dirichlet constraints (that are already in the standard basis: pointwise). Some of them are unimportant and will soon drop out, leaving only a few important ones.

1.4 Lagrange Multipliers

1.4.1 *Vector of Lagrange Multipliers*

To solve this, let's introduce a new vector of N (unknown) Lagrange multipliers:

$$\lambda \equiv (\lambda_1,\ \lambda_2,\ \lambda_3, \ldots, \lambda_N)^t \in \mathbb{R}^N.$$

This way, we actually need to solve a new $2N \times 2N$ system:

$$\begin{pmatrix} A^{-1} + A^{-t} & A^{-t} \\ A^{-1} & (0) \end{pmatrix} \begin{pmatrix} w \\ \lambda \end{pmatrix} = \begin{pmatrix} \mathbf{0} \\ b \end{pmatrix}.$$

How to solve this numerically? Recall that only A^{-1} is available, not A. So, first, solve (implicitly) for w at the bottom. Then, solve for λ at the top. (This could be viewed as an alternating block relaxation.) Then, look at your λ_n's. The smaller the $|\lambda_n|$, the weaker the nth constraint. Very weak constraints are nearly inactive and can drop out. Likewise, very strong constraints are probably noisy and should drop out too.

This leaves fewer constraints, and a smaller system, with a rectangular upper-right and lower-left blocks. To solve it numerically, restart the process all over again, and so on. Better yet, instead of alternating block relaxation, use the Schur complement (block-LU: Chapters 19 and 20 in Ref. [24]), with CGS (or QMR) on top.

(The system is symmetric but not SPD.) In the end, only a few λ_n's (active degrees of freedom) will remain, which is good for stability and denoising. Why?

To see why, look at your data again. In particular, look at two x_j's that are nearly the same (so they make similar rows in A^{-1}). But they may have very different b_j's, one of which must be noisy. Fortunately, in the above process, it will drop out, along with the jth row in the lower-left block and the jth column in the upper-right block. This will avoid a nonphysical solution (too oscillatory and unstable).

1.5 Classification Algorithm

1.5.1 *Lagrange Multipliers for Inequalities*

So far, we had Dirichlet constraints: nail $f(x_n)$ at b_n. Thanks to our algorithm, we managed to expand f as a linear combination of radial basis functions. This is quite useful: given a new $x \in \mathbb{R}^d$ (not in our data), we can now assign a proper value to it too: $f(x)$.

Let's move on to a different kind of problem: classification. Here, we need to do less: not to nail f any more but just make sure that $f(x_n)$ has the same sign as b_n. To guarantee this, require that

$$b_n f(x_n) \geq 1 \quad (1 \leq n \leq N).$$

This will indeed classify; split our x_n's into two (disjoint) subsets: those with a positive $f(x_n)$ (and b_n) and those with a negative $f(x_n)$ (and b_n). This will be quite useful: given a new $x \in \mathbb{R}^d$ (not in our data), where to throw it? Easy: calculate $f(x)$, look at its sign, and place x among friends, in the correct subset.

How to do this in a stable way? First, define a new diagonal matrix, with the b_n's on its main diagonal:

$$B \equiv \text{diag}(b_1, \; b_2, \; b_3, \ldots, b_N).$$

Next, factorize A^{-1} as a product:

$$A^{-1} = A^{-1/2} A^{-t/2},$$

where $A^{-1/2}$ is just a notation for a new matrix, with infinitely many columns:

$$A^{-1/2} \equiv (q(x_n, m))_{1 \leq n \leq N, \; 0 \leq m < \infty},$$

where, for $x \in \mathbb{R}$,

$$q(x,m) \equiv \frac{x^m}{\sqrt{m!}} \exp\left(-\frac{x^2}{2}\right)$$

and

$$A^{-t/2} \equiv \left(A^{-1/2}\right)^t .$$

Here, we mainly focus on this 1-D case: $d = 1$ and $x \in \mathbb{R}$. For $d > 1$, on the other hand, both x and m are d-dimensional:

$$x \equiv \left(x^{(1)}, x^{(2)}, \ldots, x^{(d)}\right) \quad \text{and} \quad m \equiv \left(m^{(1)}, m^{(2)}, \ldots, m^{(d)}\right),$$

so q is a bit more complicated — a tensor product over d indices: $0 \le m^{(1)}$, $m^{(2)}$, …, $m^{(d)} < \infty$ (powers of coordinates):

$$q(x,m) \equiv \exp\left(-\frac{\|x\|^2}{2}\right) \frac{x^{(1)m^{(1)}}}{\sqrt{m^{(1)}!}} \cdot \frac{x^{(2)m^{(2)}}}{\sqrt{m^{(2)}!}} \cdot \cdots \cdot \frac{x^{(d)m^{(d)}}}{\sqrt{m^{(d)}!}} .$$

Still, for simplicity, let's forget about this and focus on the 1-D case. Note also that the minus sign in $A^{-1/2}$ is just a convention. After all, our $A^{-1/2}$ is not square and has no inverse at all.

Is this decomposition useful in any way? How to handle so many columns? Well, look at what we did before. So far, we had a finite number of (unknown) coefficients:

$$w_1,\ w_2,\ w_3, \ldots, w_N .$$

These coefficients were used to write f in terms of radial basis functions (and b in terms of columns of A^{-1}). In the beginning, we did this strictly. In the end, we did this only loosely: some constraints turned out to be too weak and trivial (or too strong and noisy) and were disregarded. This gave us a new system, with more coefficients than constraints, and more stability.

Here too, we still keep the same principle: having more coefficients than constraints is good for stability. But now, the situation is even better; we seek now infinitely many coefficients:

$$w_1,\ w_2,\ w_3, \ldots.$$

These new coefficients will help expand f as an (infinite) power series: a Taylor series (around 0). First, place them in a new (infinite) column vector:

$$w \equiv (w_1,\ w_2,\ w_3, \ldots)^t.$$

How to uncover these new w_n's? Well, they will be coefficients again: not of columns of A^{-1} any more but of columns of $A^{-1/2}$. To have good stability, minimize a new kind of energy:

$$\frac{1}{2}\|w\|^2,$$

subject to our N inequalities:

$$BA^{-1/2}w \geq \mathbf{1}$$

(a column vector, containing 1's). Better yet, thanks to Lagrange's theory, we can reformulate this even better: find infinitely many coefficients

$$w \equiv (w_1,\ w_2,\ w_3, \ldots)^t$$

and N (positive or zero) Lagrange multipliers

$$\lambda \equiv (\lambda_1 \geq 0,\ \lambda_2 \geq 0, \ldots, \lambda_N \geq 0)^t$$

that minimize a new function (with a new term: $-\lambda^t$ times the original inequalities):

$$\frac{1}{2}\|w\|^2 - \lambda^t BA^{-1/2}w + \lambda^t\mathbf{1}.$$

(The minus sign comes from the "$\geq$" in our inequalities. Think of a sum of two convex functions: each pulls toward its own minimum, so the local minimum of the sum is at the tangent point in between, as in Hahn–Banach theory.) Let's differentiate this new function with respect to w and equate to 0:

$$w^t = \lambda^t BA^{-1/2}$$

or

$$w = A^{-t/2}B\lambda.$$

By substituting this in our (new) minimization problem, we can now eliminate w altogether and minimize

$$\lambda^t \mathbf{1} - \frac{1}{2}\lambda^t B A^{-1} B \lambda.$$

So, our problem got easier: we don't need to find infinitely many coefficients any more but only N Lagrange multipliers (positive or zero):

$$\lambda_n \geq 0 \quad (1 \leq n \leq N).$$

What is the meaning of such a λ_n? It tells us how powerful the nth inequality is. The bigger the λ_n, the stronger the nth inequality. If, however, we expect noise and want to force the inequality only loosely, then we could require

$$0 \leq \lambda_n \leq C$$

(for some big constant C).

Once this is solved numerically, many λ_n's will vanish: their inequalities will be satisfied automatically anyway, so there is actually no need to force them at all! This is a good sign: only a few degrees of freedom are really active: no instability or overfitting!

This is quite optimal: not too many (active) degrees of freedom and also not too few. For comparison, look at linear regression, which allows just $d+1$ degrees of freedom. These are not enough: a straight line can't classify well. Here, on the other hand, we allow more degrees of freedom: a nonlinear curve can classify better, with both stability and accuracy at the same time. (For applications and exercises, see Chapter 5.)

Chapter 2

Introduction to Quantum Mechanics from a Geometrical Point of View

To introduce quantum mechanics from scratch, we start from position–momentum. After all, these are the most elementary physical properties. In classical physics, they span the phase plane. In quantum mechanics, on the other hand, they are random variables: nondeterministic observables (measurables). Still, even in quantum mechanics, they are quite geometrical: they span the complex plane, where the Schrodinger equation makes them evolve dynamically in time.

It is good for you to start from elementary position–momentum, to gain some physical insight and intuition before diving into a deeper study [23]. After all, you want to see the entire forest, not only the individual trees. In our complex plane, we'll later introduce time–energy polar coordinates, to model the physical dynamics, and the uncertainty principle that makes the world tick.

Having said that, let's start from the beginning. How to make quantum mechanics a little more visual and geometrical? From classical physics, borrow the phase plane. Algebraically, this is also a field: the complete complex field. In it, the Cartesian coordinates have a physical meaning too: the (horizontal) real axis tells us the position of the particle (in 1-D). On it, the position is just an (uncertain) random variable (as illustrated in Figures 6.1 and 6.2): we never know its concrete value but only its distribution (its wave function, squared). Likewise, the (vertical) imaginary axis marks the momentum of the particle, distributed by the momentum wave function: the Fourier transform of the original wave function. If you like, the Fourier transform rotates the entire complex plane counterclockwise (by angle 90°). This way, it maps the real axis (with the wave function defined on it) onto the imaginary axis (with the momentum wave function defined on it).

This also makes sense analytically. Indeed, thanks to our phase plane, we can write a new complex PDE, which produces a coherent state. Later on, we'll also see its dynamics: it will rotate in the phase plane, just like the well-known (classical) harmonic oscillator. This will make quantum mechanics (and its dynamics) a little more visual and understandable.

2.1 Measurable: Observable

2.1.1 *Momentum Operator*

So far, we looked at AI and machine learning. How is this related to quantum mechanics? Well, so far, our energy operator was the Laplacian:

$$-\triangle = i\nabla^t i\nabla,$$

where ∇ is the gradient and $i = \sqrt{-1}$. Thus, minimum energy means minimum gradient (and variation), leading to a smooth function, and stability.

Upon dividing by $2m$ (where m is mass), this also takes a physical face: kinetic energy. Indeed, $i\nabla$ represents momentum. More precisely, in 1-D, momentum is the differential operator

$$P \equiv -i\bar{h}\frac{\partial}{\partial x},$$

where

$$i = \sqrt{-1}, \quad \text{and} \quad \bar{h} \doteq 10^{-34}$$

(Planck's constant). This way, the specific momentum p remains uncertain and unknown: a random variable (as discussed in Section 6.2). Indeed, better not measure p physically, or you'd change the original physical state forever, with no return. Nevertheless, p still has an (implicit) mathematical meaning: an eigenvalue of P, ready to help us statistically; calculate the expected momentum.

In P, why do we have the coefficient $i = \sqrt{-1}$? This makes P Hermitian (with respect to integration in 1-D). Thanks to this, the eigenvalue p must be real: a legitimate physical quantity. This makes P a legitimate measurable (or observable), ready to measure (or observe) a physical quantity (not quite experimentally but only mathematically and statistically).

2.1.2 *Position Operator*

Likewise, the position operator X tells us about another (elementary) physical property: the position of the particle in 1-D — a specific point $x \in \mathbb{R}$. Again, x is never measured physically but remains implicit: an uncertain random variable. How to study it statistically? Look at x as an eigenvalue of X. For this, write X in its diagonal form (as an infinite diagonal matrix):

$$X \equiv \text{diag}(x) \quad (x \in \mathbb{R}).$$

This way, each eigenvalue $x \in \mathbb{R}$ is already in its right place: on the main diagonal, ready to help calculate the expected position of the particle on the 1-D axis (see exercises in Chapter 7).

2.1.3 *Experiment: Hidden Logic?*

Look at the position x again. Is it physical at all? Not quite. After all, to uncover x, you must look (or measure, in an experiment). But then again, where to look? Usually, we use our logic and look where the particle is expected to be. But this is not quite fair. Ideally, we should look everywhere: at all points in $\mathbb{R}$ at the same time! Is this possible at all? Thus, position is more mathematical than physical: a random variable only. Fortunately, our new operator X is more global and "looks" at all x's simultaneously. The same reasoning holds for momentum too and its own operator: P. Let's study them algebraically.

2.1.4 *The Commutator*

How does this help model quantum mechanics? Well, the original numbers x and p commute:

$$xp = px.$$

But our new operators, X and P, don't: they have a nonzero commutator:

$$[X, P] \equiv XP - PX \neq 0.$$

To see this, look at a (differentiable) function

$$s : \mathbb{R} \to \mathbb{C},$$

defined for all $x \in \mathbb{R}$. How does the commutator act on it? Like this:

$$[X, P]s = (XP - PX)s = i\bar{h}((xs)' - xs') = i\bar{h}s.$$

Thus, the commutator is constant:

$$[X, P] = i\bar{h}.$$

This is quantum mechanics for you: measuring position and then momentum is not the same as measuring momentum and then position (not even statistically). Why? Because the experiment itself changes the picture.

2.2 The Fourier Transform

2.2.1 *Fourier Transform as Interference*

Assume that our function

$$s : \mathbb{R} \to \mathbb{C}$$

is not only differentiable but also (square-)integrable in $\mathbb{R}$. What is its Fourier transform? This is a new function of the new variable $p \in \mathbb{R}$:

$$(Fs)(p) \equiv \int_{-\infty}^{\infty} \exp(-ixp/\bar{h})s(x)dx.$$

Here, p is proportional to the wave number $k \in \mathbb{R}$:

$$p = k\bar{h} \quad (k, p \in \mathbb{R}).$$

This is de Broglie's law. This gives p its physical meaning: momentum. Indeed, $p = k\bar{h}$ tells us how fast the exponent cycles with x.

This is duality indeed: the particle is also a wave, oscillating (at frequency ω) along the x-axis. This cycling is in the complex plane: thanks to it, the wave oscillates like a (co)sine as it propagates. (Don't confuse this with spin, which cycles in 3-D, perpendicular to the direction of propagation.)

In the above integral, what happens physically? Interference! To see this, look at a fixed $k \in \mathbb{R}$. What is the kth wave? This is $\exp(ikx)$: an eigenfunction of P, with eigenvalue $p = k\bar{h}$. This wave (Fourier mode) will serve as a component in s (with some coefficient). What is this coefficient? This is the above integral. How is it calculated? Well, all x's contribute to it. Each x contributes a complex number (of magnitude one), known as phase: $\exp(-ikx)$, times $s(x)$. In the above integral, all these phases interfere (or superpose, or add up), to make one new complex number: our coefficient, ready to multiply the kth wave, and form a component in s. In the inverse Fourier transform, these components will superpose (or expand) s back again.

2.2.2 *Toward Quantum Field Theory*

Later on, we'll mirror this. How? We'll make three (little) changes:

- Replace position by time: t. After all, time is a random variable in its own right.
- Replace momentum by a more fundamental quantity: energy. After all, energy is a random variable in its own right.
- But energy is not continuous but only discrete: it comes in packets, or quanta. Thus, to model it, replace the Fourier transform by the Fourier series.

How to produce n quanta of energy? To this, each time t will contribute its own phase: $\exp(in\omega t)$ (times the probability to be at time t). Together, all these contributions (from all hypothetical t's) will interfere, and superpose n quanta of energy, to help jump to the nth energy level. This is quantum mechanics in a nutshell.

So far, we started from x: the position where the particle could be (at a particular time). Then, we moved to t: the time when the particle could be (at a particular x). Next, combine x with t and form the x-t system: spacetime. In it, look at a particular path, leading from point to point in spacetime (or from event to event). Physically, this draws some (hypothetical) motion in spacetime. How likely is the particle to take this path? This depends on another question: how much phase is accumulated along the path? This depends on yet another question: how much energy is accumulated along the path? Indeed, this energy is needed to "push" the particle ahead. In fact, the particle has a dual face: not only a particle but also a wave. In this sense, energy will convert into frequency, used to make phase grow, and the wave oscillate and propagate forward. This will tell us how likely the particle is to take this path.

This way, each path will have its own probability. Together, all (hypothetical) paths will superpose the path integral, telling us how likely the particle is to travel from our initial point to our final point in spacetime (on any path).

This is quantum field theory in a nutshell. We'll come back to this later. Still, for the time being, let's forget about time–energy and focus on position–momentum, and the Fourier transform that connects them.

2.2.3 *Fourier Transform as Rotation*

How to use the Fourier transform in practice? Let's use it to diagonalize the momentum operator P. For this, assume that s' is (square-) integrable too, and integrate by parts:

$$
\begin{aligned}
FPs &= -i\bar{h}\int_{-\infty}^{\infty} \exp(-ixp/\bar{h})s'(x)dx \\
&= -i\bar{h}\int_{-\infty}^{\infty} (ip/\bar{h})\exp(-ixp/\bar{h})s(x)dx \\
&= pFs.
\end{aligned}
$$

So, we actually diagonalized P:

$$P = F^{-1}pF.$$

How does an eigenfunction of P look like? Algebraically, this is a column in F^{-1}. Physically, on the other hand, this is the kth wave: $\exp(ikx)$ (eigenvalue: $p = k\bar{h}$).

How does the Fourier transform look like? This is illustrated geometrically in the complex plane (Figure 2.1). x is the horizontal (real)

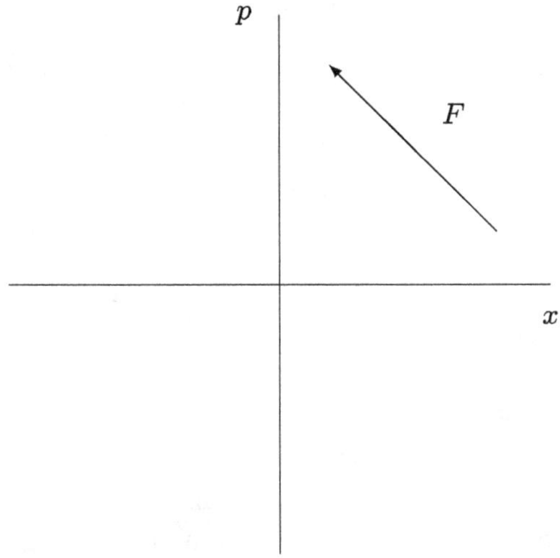

Figure 2.1. The Fourier transform F rotates counterclockwise. This way, you have two (equivalent) options: either apply P horizontally and then rotate, or rotate first, and then apply p vertically.

axis, and p is the vertical (imaginary) axis. This way, our function $s \equiv s(x)$ is defined on a horizontal line. F rotates counterclockwise (like multiplying by i) and defines Fs on the corresponding vertical line. This way, thanks to its diagonal form, P is invariant under F. Indeed, to use P, you have two (equivalent) options: either apply P horizontally and then let F rotate, or work the other way around; let F rotate first, and then multiply by p on the vertical line.

2.3 Wave Function

2.3.1 *Wave Function*

How to use this in practice? Let

$$u : \mathbb{R} \to \mathbb{C}$$

be the wave function, defined for all $x \in \mathbb{R}$. What is its physical meaning? To see this, look at $|u|^2$. Once normalized (to have integral 1 in $\mathbb{R}$), $|u|^2$ will be the probability to be at x.

2.3.2 *Momentum Wave Function*

Moreover, thanks to u, we'll also design the momentum wave function: Fu. Still, this is not quite physical or experimental but more theoretical and logical: it comes from mathematical physics.

In classical physics, the particle can only be at one place at a time. In quantum mechanics, on the other hand, it can be at all x's at the same time (each at its own probability). Now, the Fourier transform is a function of p, not x. For a fixed p, what is this function? It is an integral over all x's. How to calculate it? We already know how: each x contributes a complex number of magnitude 1: $\exp(-ixp/\bar{h})$. This is called a phase. Each phase is then multiplied by the corresponding $u(x)$ and contributes to the integral. This is how all these contributions superpose the momentum wave function: $(Fu)(p)$. Once normalized (to have integral 1 on the entire vertical line), $|Fu|^2$ will be the probability to have momentum p. Later on, we'll see a few interesting examples.

In vacuum, for example, both u and Fu will be the same: a real Gaussian (centered at 0). This is quite special and unique: a static state, of constant (and minimal) energy. But this is not the only possible state: the particle can also have more energy. Moreover, the particle can even take a more dynamic state. In a coherent state, for example, both u and Fu will change in time: u will be a complex Gaussian (defined on each horizontal line), and Fu will mirror it on the corresponding vertical line. This picture will not be static but dynamic: it will travel and rotate, as a whole. This is why the coherent state is so realistic: it will model the familiar harmonic oscillator, well known in classical mechanics.

2.3.3 *Complex Wave Function*

So far, our wave function u is defined in $\mathbb{R}$ only. Let's extend it and define it in $\mathbb{C}$ as well. This will be our new complex wave function. Once restricted to a horizontal line, it will reproduce u back again. Moreover, once restricted to a vertical line, it will give us the momentum wave function: Fu. Let's see a simple example: a degenerate state, obtained after measurement.

In our complex plane (Figure 2.1), design a new complex function (defined for all $z \in \mathbb{C}$):

$$\exp\left(\frac{z^2 - \bar{z}^2}{4\bar{h}}\right) = \exp\left(i \cdot \Re z \cdot \Im z/\bar{h}\right) = \exp(ixp/\bar{h}).$$

Is this familiar? This is actually the inverse Fourier transform. In this simple example, this will be our complex wave function. But this is quite degenerate: it models not the original physical state but only a new degenerate state: after measurement. To see this, assume that we already measured momentum and discovered its deterministic value: $p_0 \in \mathbb{R}$. Geometrically, we are now confined to one horizontal line (with a fixed imaginary part):

$$p = \Im z = p_0.$$

Once restricted to this line, our complex wave function makes a new (degenerate) wave function (containing just one Fourier mode):

$$u(x) = \exp(ixp_0/\bar{h}).$$

Is this familiar? This is an eigenfunction of P, with eigenvalue p_0. As x changes, this can only change phase (angle), not magnitude. In fact,

$$|u|^2 \equiv 1.$$

What does this mean physically? The particle is equally likely to be anywhere in $\mathbb{R}$. This is the uncertainty principle: once we know momentum, we know nothing about position any more! (Too late to measure position now: this will only tell us where the particle is now, not where it was in the beginning.) After all, to measure momentum, we already made an experiment, which changed our original u completely, losing many interesting Fourier modes, along with their valuable physical information. This is the second law of thermodynamics: entropy must grow, and information must get lost. How to avoid this? Don't measure, and don't spoil your original wave function!

2.4 Harmonic Oscillator and Its Hamiltonian

2.4.1 *Harmonic Oscillator and Its Hamiltonian*

Look at Planck's constant again: $\bar{h}$. What physical unit does it have? A product: length times momentum. This way, $xp/\bar{h}$ is a pure number (with no physical unit at all), ready to be plugged into the exponent function (as we did above). This is also why our momentum operator

$$P = -i\bar{h}\frac{\partial}{\partial x}$$

has the correct unit: momentum.

Consider now a particle of mass m, and frequency ω (of unit 1/time). What is its phase plane? This is a new complex plane: the real coordinate is x (unit: length), and the imaginary coordinate is $p/(m\omega)$ (unit: length too). In this plane, our particle behaves like a harmonic oscillator. Energy is a sum of two parts: potential energy, plus kinetic energy. This makes a new operator — the Hamiltonian:

$$H \equiv \frac{m\omega^2}{2}\left(X^2 + \frac{1}{m^2\omega^2}P^2\right).$$

This is a good definition; both terms have the same unit: energy (mass times velocity squared). In this phase plane, H will help draw energy levels: circles. On each circle, energy will be constant, and position–momentum will oscillate (precess). This will give the particle its dual face: not only a particle but also a wave.

2.4.2 *Planck: Energy Is Frequency!*

In quantum mechanics, position and momentum don't commute any more: they have a nonzero commutator. What does this mean physically? Order matters: measuring position and then momentum is not the same as measuring momentum and then position (not even statistically). For this reason, measuring position–momentum simultaneously is just mathematical and theoretical but not physical or practical at all. This is not because we are not quick enough to measure both at the same time but because nature is inherently stochastic and nondeterministic. In nature, each physical property is uncertain: a random variable. For example, position is never measured physically but only characterized by the wave function. This allows many hypothetical positions at the same time (each at its own probability). On top of this, we can even add (superpose) a few wave functions and obtain a yet richer state. For this, use linear algebra only: no need to make any experiment at all. Don't measure, and don't spoil your wave function! The nonzero commutator will soon help design a new (linear) relation between energy and frequency. This is Planck's law, which helped explain a few mysterious phenomena in physics.

Look at our operators again: X and P. Together, they form a new complex operator (and its complex conjugate), ready to help factorize the Hamiltonian:

$$\begin{aligned}
&\frac{m\omega^2}{2}\left(X - \frac{i}{m\omega}P\right)\left(X + \frac{i}{m\omega}P\right)\\
&= \frac{m\omega^2}{2}\left(X^2 + \frac{1}{(m\omega)^2}P^2 + \frac{i}{m\omega}[X,P]\right)\\
&= H - \frac{\omega\bar{h}}{2}.
\end{aligned}$$

Here, thanks to the commutator, we got a new term (linear in ω): minimal energy. This is indeed Planck's law: energy is (proportional to) frequency.

2.4.3 *Hamiltonian and Its Factorization*

What do we have in this formula? On the left-hand side, we have a product of two factors (in round parentheses). But we can also change this a little: multiply the former factor by $-i$ and the latter by i. After all, this changes nothing. As a result, we'll get

$$\begin{aligned}
H &= \frac{m\omega^2}{2}\left(X - \frac{i}{m\omega}P\right)\left(X + \frac{i}{m\omega}P\right) + \frac{\omega\bar{h}}{2}\\
&= \frac{m\omega^2}{2}\left(-\frac{1}{m\omega}P - iX\right)\left(-\frac{1}{m\omega}P + iX\right) + \frac{\omega\bar{h}}{2}.
\end{aligned}$$

Is this familiar? This is actually the Fourier transform: F. Indeed, geometrically, F rotates counterclockwise (Figure 2.1). This way, X takes the role of $P/(m\omega)$ (which, in its diagonal form, acts upward), and $-P/(m\omega)$ takes the role of X (which acts rightward). This is also apparent from our new factors: $-P/(m\omega)$ comes now first (in the real part) and X follows (in the imaginary part).

This will also make sense algebraically geometrically. On the vertical axis (in terms of momentum), our operators will take a new face. X will differentiate (upward, not rightward), and P will become diagonal (upward, not rightward). On their way (to complete 180°), they will also pick a mines sign. Why? Because this is the Fourier transform (twice), not its inverse!

How to see this more explicitly? On the vertical axis, define X and P in such a way that they commute with F:

$$P \equiv \operatorname{diag}(p) \quad \text{and} \quad X \equiv i\bar{h}\frac{\partial}{\partial p}.$$

This way, P indeed commutes with F:

$$PF = pF = FP.$$

Likewise, X commutes with F too:

$$\begin{aligned} XFs &= i\bar{h}\frac{\partial}{\partial p}\int_{-\infty}^{\infty} \exp(-ixp/\bar{h})s(x)dx \\ &= i\bar{h}(-i/\bar{h})\int_{-\infty}^{\infty} \exp(-ixp/\bar{h})xs(x)dx \\ &= FXs. \end{aligned}$$

Later on, our new decomposition of the Hamiltonian will also take a more physical face: evolution and dynamics (Sections 3.5.1 and 3.5.2). For this, we'll multiply the original factors by not only $\pm i$ but also $\exp(\pm i\omega t)$. This will keep the Hamiltonian unchanged. This is the first law of thermodynamics, in its stochastic face: energy will remain conserved, all the time (unchanged under precession, or circular motion).

Which decomposition is better? The former or the latter? Better stick to the former. The original factors avoid the minus sign and are more useful and popular.

2.4.4 *Energy Level: Ground State*

In the harmonic oscillator, the most elementary state is the ground state (also known as zero-point). What is this? Mathematically, this is a function in the null space of $X + iP/(m\omega)$. (In our original decomposition of the Hamiltonian, this is the second factor.) Thus, this is an eigenfunction of H, with the minimal eigenvalue: $\omega\bar{h}/2$.

Physically, on the other hand, the ground state represents vacuum, with minimal energy: $\omega\bar{h}/2$. This is vacuum energy, coming from virtual particles that exist and don't exist at the same time.

But wait a minute: what do you mean by vacuum? Where is our original particle? Doesn't it exist any more? This is a good question, which deserves two answers: a physical one and a mathematical one:

- In quantum mechanics, nothing really exists! The particle needs some energy, to support it and let it "exist." At some probability, the energy is minimal: $\omega\bar{h}/2$ only. This is not enough to support a particle. Still, nature has a state like this: the ground state.
- Every state of constant energy (including the ground state) is more mathematical than physical. After all, we never measure energy and never get to know what it really is!

How does the ground state look like? In other words, what is its wave function (that encapsulates its physical properties)? Well, it looks like a real Gaussian (centered at 0). Can it evolve in time? Not much. It can only precess: cycle (at frequency $-\omega/2$). For this, it must contain a time factor: $\exp(-i\omega t/2)$. But this has little physical effect. In particular, energy remains $\omega\bar{h}/2$, all the time. Stochastically, position-momentum remain the same too, all the time.

$\omega\bar{h}/2$ is the minimal energy. But this is not the only possible energy. In fact, H has many more eigenfunctions, with bigger and bigger eigenvalues (energy levels). Thanks to their higher energy, they precess at a higher frequency:

$$-\left(n + \frac{1}{2}\right)\omega, \quad n = 0, 1, 2, 3, \ldots .$$

Again, this precession has little physical effect. In particular, energy remains the same, all the time. Stochastically, position–momentum remain the same too: in one eigenfunction, the probability to be at a particular position (and to have a particular momentum) remains the same too, all the time. After all, the (momentum) wave function can only change phase (angle), not magnitude. This has no effect on the probability at all.

Still, energy is superior to position–momentum. In what sense? Well, position–momentum remain unchanged only in one eigenfunction, not in a sum of two or more. Energy, on the other hand, is more robust: it remains unchanged even in a Fourier series: an infinite sum of eigenfunctions. In what sense? In the stochastic sense: it keeps the same probability to be at each energy level. Why is energy so stable and robust?

In nature, energy is most elementary. Indeed, the most fundamental law is conservation of energy. This is the first law of thermodynamics. Here, in quantum mechanics, we see this law in its nondeterministic face: constant probability to be at a particular energy level (spanned by one particular eigenfunction of H). Position–momentum, on the other hand, are not so robust: they remain unchanged only in one eigenfunction, not in a sum of two or more (not even nondeterministically): the probability to be at a particular position (and to have a particular momentum) does change dynamically over time (unless the state is just one eigenfunction). Why?

Each eigenfunction is special: it preserves energy and forms one energy level. Still, we can also design a more interesting state, with a richer dynamics: a superposition (sum) of a few eigenfunctions. (For example: a coherent state.) Here, phase matters, and has a real physical effect: not on energy but on position–momentum. Indeed, different eigenfunctions precess at different frequencies, interfere at a new proportion, and produce a dynamic wave function, evolving all the time (in not only phase but also magnitude). Physically, this will produce beautiful strips of light (as in the double-slit experiment: Figure 4.1).

A good example is the coherent state. How does it look like? Initially, it looks like a complex Gaussian (applied to $\bar{z}$, and shifted in the complex plane). But this is not static, but dynamic, and will soon change (in terms of position–momentum, not energy). As time goes by, different components in it will precess at different frequencies and interfere at a new proportion.

This will be our complex wave function, defined in $\mathbb{C}$ (not just in $\mathbb{R}$). What does it tell us physically? To see this, restrict it to some horizontal line. This will be the wave function. How to use it in practice? Do three things:

- calculate its absolute value,
- square it up, and
- normalize, to have integral 1 (on the entire line).

This will be the probability to be at x. The same works with momentum too (in terms of p, on the corresponding vertical line). This is how position–momentum mirror one another and are treated in a uniform way.

And what about energy? To have its probability too, look at all energy levels (indexed by $n = 0, 1, 2, 3, \ldots$). This will make a new discrete wave function: an infinite sequence of complex numbers (a function of n only). How to use it? Again, do three things:

- take its absolute value,
- square it up, and
- normalize, to have sum 1 (over all n's).

This will make a new sequence of positive numbers (that sum to 1). It will serve as discrete probability: a function of n only. (See two simple examples in Figures 6.1 and 6.2.) It will answer a more fundamental question: how likely is the particle to be at the nth energy level? To answer this, we need the Poisson distribution. To design it, we must first study the complex Gaussian and expand its Fourier series.

2.5 The Coherent State

2.5.1 *Complex Gaussian and Its Fourier Series*

In our Green function (the real Gaussian G: Section 1.2.2), look at the 1-D case: $d = 1$ and $x \in \mathbb{R}$. This way, in the exponent function used there, one can actually drop the norm signs and replace them with round parentheses: $(x - \tilde{x})^2$ rather than $\|x - \tilde{x}\|^2$. Why is this good? Well, this will also inherit to the function g, obtained from G

(Section 1.2.3). It will now look like this:

$$g(x, \tilde{x}) = \exp\left(-\frac{(x-\tilde{x})^2}{2}\right)$$

(Figure 9.1). We can now extend g into a new (complex) version (defined for $z, \mu \in \mathbb{C}$):

$$\tilde{g}(z, \mu) \equiv g\left(\sqrt{\frac{m\omega}{\hbar}}z, \sqrt{\frac{m\omega}{\hbar}}\mu\right).$$

Is this a good definition? Well, in our phase plane, both z and μ have the same physical unit: length. Once multiplied by the new coefficient $\sqrt{m\omega/\hbar}$, they convert into a new pure number (with no physical unit at all), ready to be plugged into the exponent function. Likewise, the function q (Section 1.5.1) takes a new (complex) version too:

$$\tilde{q}(z, n) \equiv q\left(\sqrt{\frac{m\omega}{\hbar}}z, n\right) = \frac{\left(\sqrt{\frac{m\omega}{\hbar}}z\right)^n}{n!}\exp\left(-\frac{m\omega}{2\hbar}z^2\right)$$

($z \in \mathbb{C}$, $0 \le n < \infty$). From this, we can now design new infinite matrices (whose names will be discussed soon). First, define a new $\mathbb{C} \times \mathbb{C}$ matrix:

$$A^{-1} \equiv (\tilde{g}(\bar{z}, \mu))_{z,\mu \in \mathbb{C}}.$$

This is a good matrix: Hermitian and positive definite. Next, define also a new $\mathbb{C} \times \mathbb{N}$ matrix:

$$A^{-1/2} \equiv (\tilde{q}(\bar{z}, n))_{z \in \mathbb{C}, 0 \le n < \infty}.$$

In both matrices, in their names, there is a minus sign. This is just a convention. After all, they are not the inverse of anything. Still, they factorize as

$$A^{-1} = A^{-1/2}A^{-*/2},$$

where the superscript "*" stands for the Hermitian adjoint:

$$A^{-*/2} \equiv \left(A^{-1/2}\right)^*.$$

This way, the μth column in A^{-1} is a superposition (linear combination) of columns in $A^{-1/2}$, with coefficients from the μth column in $A^{-*/2}$. In a sense, this is a Fourier series: sum of $\bar{z}^n$'s (times their corresponding coefficients). In this sum, each $\bar{z}^n$ will make an angular Fourier mode, cycling in the complex plane, faster and faster, with more and more energy as n grows. Together, all the coefficients will yield a new discrete distribution (over $0 \le n < \infty$).

2.5.2 *Discrete Wave Function*

How to design the discrete distribution? For this, we must first have the discrete wave function: an infinite sequence of complex numbers (a function of n only). Fortunately, we already have it: the μth column in $A^{-*/2}$. But there is a problem. So far, the $\bar{z}^n$'s were normalized by $\sqrt{n!}$ only. This is not enough: to mirror the eigenfunctions of the Hamiltonian, they must also be divided by $2^{n/2}$. How to do this? In A^{-1}, look at the (z, μ)th element. Model it by another element: the $(z, \mu/\sqrt{2})$th element. How to calculate it? For this, look at the zth row in $A^{-1/2}$ and the $\mu/\sqrt{2}$th column in $A^{-*/2}$. Multiply them, element by element, and sum up. Thus, the latter column contains the correct discrete wave function.

2.5.3 *Discrete Poisson Distribution*

How to use it in practice? How to convert it into a legitimate distribution? For this, do three things:

- take its absolute value,
- square it up, and
- normalize (to sum to 1, over all the n's).

This will yield

$$\exp\left(-\frac{m\omega}{2\hbar}|\mu|^2\right)\frac{\left(\frac{m\omega}{2\hbar}|\mu|^2\right)^n}{n!}.$$

This is the Poisson distribution (Figure 9.3): an infinite sequence of positive numbers (that sum to 1), telling us the probability to be at the nth energy level, with energy $(n + 1/2)\omega\hbar$. Later on, we'll see that this is static, not dynamic: the same probability will hold, all the time. What changes dynamically is not the probability to have a particular energy but only the probability to be at x (on the horizontal line), or to have momentum p (on the vertical line). This is thanks to interference. But interference can only mix position or momentum, not energy levels that never interact.

This is why energy is more fundamental: it is more robust, stable, and reliable. On each energy level, the complex wave function can only precess (cycle) like $\bar{z}^n$, keeping the same probability to be at this energy level (as in the Poisson distribution). This is conservation of energy (in its stochastic form): same probabilities (discrete distribution), all the time.

2.5.4 *Saturated State*

Look at the complex Gaussian again: $\tilde{g}(z, \mu)$. Around a fixed μ, it is symmetric (Figure 2.2): z reflects symmetrically across μ. Indeed, the complex Gaussian is invariant (unchanged) under the transformation

$$z = \mu - (\mu - z) \to \mu + (\mu - z) = 2\mu - z.$$

Thus, in our complex plane, the complex Gaussian will make a saturated state: each vertical line will mirror a horizontal line, to tell us the probability to have momentum p (not ip).

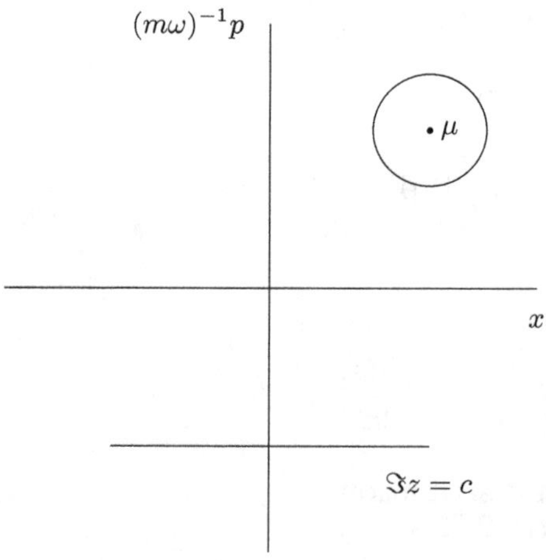

Figure 2.2. The coherent state is obtained from the complex Gaussian (applied to $\bar{z}$ and centered at μ). This is saturated: on each horizontal line, it looks the same as on the corresponding vertical line.

How to see this analytically? Later on, the complex Gaussian will solve a complex PDE, containing two PDEs. The first one differentiates horizontally: with respect to $x = \Re z$. The second one, on the other hand, differentiates vertically: with respect to $ip/(m\omega)$. Thanks to the Fourier transform, this i will soon disappear. After all, the Fourier transform rotates, so there is no need to multiply by i any more.

The first PDE will tell us how the wave function looks like in terms of x (on each horizontal line). This is invariant: apply the Fourier transform F, and you'll get the second PDE, which has the same form, and tells us how the momentum wave function looks like (in terms of p, not ip). In fact, on the corresponding vertical line, this is F times the original wave function (on the original horizontal line).

2.5.5 *De Broglie: Momentum Is Wave Number!*

How to see this geometrically? In Figure 2.2, look at a horizontal line (with a fixed imaginary part):

$$\Im z = c$$

or

$$z = x + ic$$

(for some constant $c \in \mathbb{R}$). What is c physically? Well, in the complex Gaussian, c will produce an extra oscillation: $\exp(-im\omega xc/\bar{h})$. This will shift the Fourier transform and contribute $-m\omega c$ to the expected momentum. (Don't worry: in the end, to model the coherent state itself, we'll substitute $\bar{z}$ for z, so this minus sign will disappear.) After all, the momentum is proportional to the wave number k:

$$p = k\bar{h} \quad (k, p \in \mathbb{R}).$$

This is de Broglie's law. Originally, it was introduced to prove that the electron is not only a particle but also a wave. Later on, this was also extended to any other particle as well. It tells us that, for a moderate $|k|$, $|p|$ is very small. This is why it is so important to get p centered properly: around its expected value. Thanks to our phase plane, we'll be able to do this.

On the above horizontal line, the complex Gaussian will help design the wave function of the coherent state (whose square is the probability to be at $x = \Re z$). This is saturated: it looks the same as on the corresponding vertical line. To see this, look at a few algebraic mirror transformations:

- How to change sign in the imaginary part? Interchange
$$z \leftrightarrow \bar{z}.$$
- How to change sign in the real part? Interchange
$$z \leftrightarrow -\bar{z}.$$
- How to interchange real and imaginary axes? Use two steps: first, change sign in the imaginary part. Then, rotate counterclockwise:
$$z \to \bar{z} \to i\bar{z}.$$
Or, if you like, use two other steps: first, change sign in the real part. Then, rotate clockwise:
$$z \to -\bar{z} \to -i(-\bar{z}) = i\bar{z}.$$
Either way, you actually do the same:
$$z \leftrightarrow i\bar{z}.$$

The latter transformation will be particularly useful in the following. Still, under it, the complex Gaussian is not quite invariant (in the usual sense): it does change. How to improve on this and avoid any change? On our horizontal line, "normalize": divide the complex Gaussian by the (real) factor

$$\exp\left(\frac{m\omega}{\hbar}|\Im\mu - c|^2\right).$$

This would yield a more symmetric function: invariant under interchanging real and imaginary axes. (In a sense, this makes a new hyperbolic metric.) Still, let's leave this to the exercises. After all, the complex Gaussian is good enough for us: it is invariant in the sense discussed in the following.

2.5.6 *Coherent State and Its Eigenvalue*

On our horizontal line, the coherent state will be modeled by its complex wave function: $\tilde{g}(\bar{z}, \mu)$. This is the complex Gaussian at $\bar{z}$ (rather than z). This will be our complex wave function. Once restricted to the horizontal line, it will make the wave function (in terms of x). Once restricted to a vertical line, on the other hand, it will make the momentum wave function (in terms of p, not ip). This is good: position–momentum will mirror one another and be treated symmetrically.

In our coherent state, how to design the complex wave function? For this, we must first study the original complex Gaussian: $\tilde{g}(z, \mu)$ (with no complex conjugate). In it, it will be easy enough to plug $\bar{z}$ later (rather than z) and differentiate at the new point $\bar{z}$, along its own horizontal line:

$$\bar{z} \in \{z \mid \Im z = -c\}.$$

Better yet, work with $\tilde{g}(z, \bar{\mu})$. This is the complex conjugate of $\tilde{g}(\bar{z}, \mu)$. Physically, they are the same.

Anyway, for simplicity, let's focus on the original complex Gaussian: $\tilde{g}(z, \mu)$ (with no complex conjugate at all). It solves a new complex PDE:

$$\left(z + \frac{\bar{h}}{m\omega} \cdot \frac{\partial}{\partial z}\right) \tilde{g} = \mu \tilde{g}.$$

This differentiates with respect to the complex differential

$$dz = dx + i\frac{dp}{m\omega}.$$

Thus, our complex PDE actually splits into two PDEs: one differentiates horizontally and the other vertically:

$$\left(x + i\frac{p}{m\omega} + \frac{\bar{h}}{m\omega} \cdot \frac{\partial}{\partial x}\right) \tilde{g} = \mu \tilde{g},$$

$$\left(x + i\frac{p}{m\omega} + \frac{\bar{h}}{m\omega} \cdot \frac{\partial}{\partial ip/(m\omega)}\right) \tilde{g}$$

$$= \left(x + i\frac{p}{m\omega} - i\frac{\bar{h}}{m\omega} \cdot \frac{\partial}{\partial p/(m\omega)}\right) \tilde{g} = \mu \tilde{g}.$$

Thanks to the Fourier transform, this minus sign will soon disappear. Why? Because the Fourier transform rotates in the complex plane (just like multiplying by i), so there is no need to multiply by i on top any more. How to see this? On our horizontal line, rewrite the original complex PDE as

$$\left(z - ic + \frac{\bar{h}}{m\omega} \cdot \frac{\partial}{\partial z}\right) \tilde{g} = (\mu - ic)\tilde{g}.$$

To obtain the first PDE back again, differentiate horizontally (with respect to $x = \Re z$):

$$\begin{aligned}\left(X + \frac{i}{m\omega} P\right) \tilde{g} &= \left(x + \frac{\bar{h}}{m\omega} \cdot \frac{\partial}{\partial x}\right) \tilde{g} \\ &= \left(z - ic + \frac{\bar{h}}{m\omega} \cdot \frac{\partial}{\partial z}\right) \tilde{g} \\ &= (\mu - ic)\tilde{g}.\end{aligned}$$

Algebraically, this means that $\tilde{g}$ is an eigenfunction of $X + iP/(m\omega)$ (with eigenvalue $\mu - ic$). Physically, on the other hand, this will yield the wave function on our horizontal line.

To this equation, apply the Fourier transform: F. Fortunately, F is linear, and commutes with both X and P (Section 2.4.3). Thus,

$$\begin{aligned}(\mu - ic)F\tilde{g} &= F(\mu - ic)\tilde{g} \\ &= F\left(X + \frac{i}{m\omega} P\right) \tilde{g} \\ &= \left(X + \frac{i}{m\omega} P\right) F\tilde{g} \\ &= \left(i\bar{h}\frac{\partial}{\partial p} + \frac{i}{m\omega} p\right) F\tilde{g} \\ &= i\left(\frac{\bar{h}}{m\omega} \cdot \frac{\partial}{\partial p/(m\omega)} + \frac{p}{m\omega}\right) F\tilde{g}.\end{aligned}$$

This is the same differential operator as before, except for two changes:

- A new variable: $p/(m\omega)$ (rather than x).
- A new coefficient: i.

Where did this new i come from? Algebraically, it came from the second decomposition of the Hamiltonian (Section 2.4.3). Analytically, on the other hand, it came from this: in our original complex PDE, we differentiated with respect to $ip/(m\omega)$, not $p/(m\omega)$. Physically, however, we want the momentum wave function (in terms of p, not ip). Fortunately, the Fourier transform already rotated in the complex plane, avoiding the need to multiply by i. Thanks to this, we can now solve for the momentum wave function: $F\tilde{g}$. For this, divide the above equation by i:

$$\frac{p}{m\omega}F\tilde{g} + \frac{\bar{h}}{m\omega} \cdot \frac{\partial}{\partial p/(m\omega)}F\tilde{g} = (-i\mu - c)F\tilde{g}.$$

This is our second PDE back again, except for three improvements (on the left-hand side):

- No i any more.
- No ip any more but only p.
- No minus sign and no hyperbolic geometry any more.

This is invariant: has the same form as the first PDE, except for a new variable: $p/(m\omega)$ (rather than x). How to solve it? Look at a new vertical line (with a fixed real part):

$$x = \Re z = -c$$

or

$$z = -c + i\frac{p}{m\omega}$$

(Figure 2.3). It mirrors our original horizontal line and can be mapped onto it by rotation:

$$z \to -iz.$$

On this vertical line, how to solve for $F\tilde{g}$ (to obtain the momentum wave function, in terms of p)? No need to apply the Fourier transform explicitly. Instead, do four things:

1. Look at a z on this vertical line (Figure 2.3). Rotate clockwise:

$$z \to -iz.$$

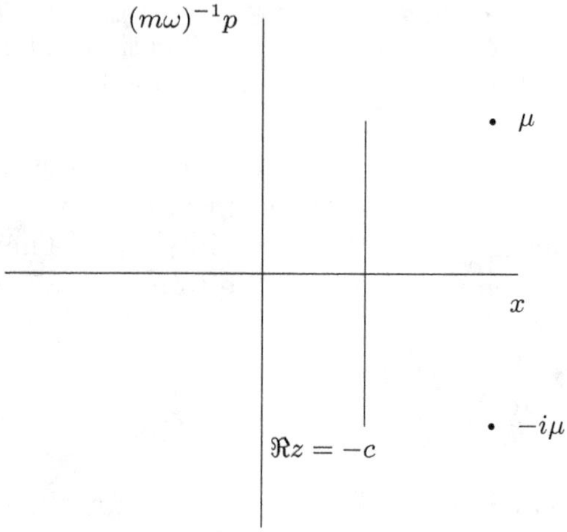

Figure 2.3. The new vertical line mirrors the original horizontal line by rotation. Indeed, rotate clockwise: $z \to -iz$, and obtain the original horizontal line back again.

2. This way, $-iz$ is on the original horizontal line. Apply $\tilde{g}$ to it.
3. On our vertical line, this is actually $\tilde{g}(-iz, -i\mu)$.
4. This is how $F\tilde{g}$ is defined on our vertical line.

Later on, we'll extend this rotation yet more and see its dynamic face too (Sections 3.5.1 and 3.5.2).

2.5.7 *Geometrical-Physical Picture*

This is analytic. But now, we want more: a geometrical-physical picture too. To see this, look at our horizontal line again (Figure 2.2). It is characterized by c: its height in the plane. What is c physically? It tells us about momentum: the bigger the $\Im\mu - c$, the bigger the expected momentum (thanks to the bigger shift in the Fourier transform). More precisely, to obtain the coherent state $\tilde{g}(\bar{z}, \mu)$, substitute $\bar{z}$ for z, so this will become $\Im\mu + c$, as required (rather than $\Im\mu - c$).

This is about momentum. But what about position? How likely is the particle to be at $x = \Re z$? The probability for this is $|\tilde{g}(\bar{z}, \mu)|^2$ (normalized to have integral 1 on the entire horizontal line). As a matter of fact, on all horizontal lines, this will give the same distribution (probability density): a real Gaussian (centered at $\Re\mu$:

Figure 9.2). Thus, in the coherent state, position and momentum are independent (but still entangled: Section 6.8.3).

Next, mirror this vertically as well: for a fixed x, look at the vertical, issuing from x. On it, thanks to our second PDE, we'll have the momentum wave function, telling us how likely the particle is to have momentum p. For this, don't forget to substitute $\bar{z}$ for z and to work with $\tilde{g}(\bar{z}, \mu)$, not $\tilde{g}(z, \mu)$. On the vertical line in Figure 2.3, for example, you'll then have expected position $\Re\mu - c$, as required (rather than $\Re\mu + c$).

This is quite uniform: the same complex wave function $\tilde{g}(\bar{z}, \mu)$ can be projected onto any (horizontal or vertical) line, and model position or momentum (in their nondeterministic face). How to use this to measure? In $\tilde{g}(\bar{z}, \mu)$, disregard magnitude altogether and focus on phase only (a degenerate state: Section 2.3.3):

$$\exp\left(-i\frac{m\omega}{\bar{h}}\Re(\bar{z}-\mu)\Im(\bar{z}-\mu)\right) = \exp\left(i\frac{m\omega}{\bar{h}}\left(\Re z - \Re\mu\right)\left(\Im z + \Im\mu\right)\right).$$

On our original horizontal line, how to measure momentum? Do three things:

- Set $\Im z = c$.
- To this function, apply $P = -i\bar{h}\partial/\partial x$.
- This will pick $m\omega(\Im\mu + c)$. This agrees with the expected momentum, discussed above.

On our vertical line, on the other hand, how to measure position? Do three other things:

- Set $\Re z = c$.
- To the above function, apply $X = i\bar{h}\partial/\partial p$.
- This will pick $\Re\mu - c$. This agrees with the expected position, discussed above.

Here, x picks a minus sign. In our former model (Section 2.5.6), on the other hand, there is no need: this is already built in, more elegantly and implicitly. Indeed, after the rotation, x points downward, not upward!

So far, we studied position–momentum and treated them uniformly. But now, we want even more: to expand the coherent state in terms of not only x or p but even n: energy levels. Fortunately, we already did this: the discrete wave function is already stored in the $\mu/\sqrt{2}$th column in $A^{-*/2}$. We even did more: we already squared it up and normalized (over all n's). This gave us the Poisson distribution: the probability to be at the nth energy level, with energy $\omega\bar{h}(n+1/2)$. Physically, as n grows, $\bar{z}^n$ precesses (cycles) faster and faster, with more and more energy. How to see this geometrically?

2.5.8 *Schrodinger's Equation*

In our phase plane, our operators have two different forms: horizontal and vertical. How do they look like? Horizontally, X is diagonal:

$$X = \text{diag}(x).$$

This way, each eigenvalue $x \in \mathbb{R}$ corresponds to one eigenfunction: a delta function (centered at x). Vertically, on the other hand, X differentiates:

$$X = i\bar{h}\frac{\partial}{\partial p}.$$

This way, each eigenvalue $x \in \mathbb{R}$ corresponds to a new eigenfunction: the Fourier mode $\exp(-ixp/\bar{h})$ (as a function of p). This is the dual face of nature: not only a local particle (at x) but also a global wave (in p).

Likewise, P shares the same structure (only the other way around): vertically, it is diagonal:

$$P = \text{diag}(p).$$

This way, each eigenvalue $p \in \mathbb{R}$ corresponds to one eigenfunction: a delta function (centered at p). Horizontally, on the other hand, P differentiates:

$$P = -i\bar{h}\frac{\partial}{\partial x}.$$

This way, each eigenvalue $p \in \mathbb{R}$ corresponds to a new eigenfunction: the Fourier mode $\exp(ixp/\bar{h})$ (as a function of x). And what about H? How does it look like geometrically? On each radial ray (issuing from the origin), H could be illustrated as diagonal:

$$H = \text{diag}(H),$$

where H stands for not only the Hamiltonian but also its eigenvalue: energy. For example, let's start from the horizontal ray, pointing rightward. On it, the new coordinate H is not the same as x:

- Unlike x, H is not continuous but only discrete:

$$H = \omega\bar{h}\left(n + \frac{1}{2}\right), \quad n = 0, 1, 2, 3, \ldots.$$

- Unlike x, H (and n) grow not only linearly but also quadratically (like x^2).

And what are the eigenfunctions? How do they look like? On the ray, they are quite local and schematic: a delta function (centered at a particular H). In terms of z, on the other hand, they will look more global and physical: $\bar{z}^n$ (times the ground state). Better yet, on the real axis, they will look like a polynomial of degree n in x (times the ground state).

Fortunately, the eigenfunctions will remain static: as time goes by, the entire ray will rotate (counterclockwise), carrying the same eigenfunctions (and eigenvalues) with it, round and round. This means that time grows angularly (circularly, counterclockwise, round and round). This way, time will supply the missing polar coordinate (perpendicular to the ray). In this direction, the Hamiltonian will differentiate:

$$H = i\bar{h}\frac{\partial}{\partial t}.$$

This is Schrodinger's equation. Thanks to it, H shares the same dual structure as X and P (only with a little more complicated eigenfunctions, to be discussed soon). This is the dual face of the particle: not only a local quantum of energy but also a global wave, propagating in time. (For exercises, see Chapter 7.)

Chapter 3

Dynamics: Time–Energy Polar Coordinates

By now, our (complex) phase plane is spanned in its Cartesian coordinates: the real axis marks the position of the particle and the imaginary axis its momentum. Better yet, we can also draw polar coordinates: energy makes the radius, and time the angle. More precisely, energy is discrete: start from the origin and draw bigger and bigger circles (each with its own constant energy). These are the energy levels: only these are allowed. This is quantum mechanics for you: energy comes in quanta — discrete packets.

On each circle (or energy level), energy is constant. As time goes by, we rotate (or precess) on the circle. The higher the energy, the bigger the circle, and the faster the precession. This is conservation of energy: we always remain on the same circle, with the same energy, all the time.

Thus, this precession is only mathematical: it has no physical effect. Even in a superposition (sum) of two states, energy is still conserved (in a stochastic sense): it keeps the same probability to be on either of these two circles, all the time. Position–momentum, on the other hand, remain the same only if we are confined to one circle. In a superposition over two circles (or more), we'll have a dynamical interference. Later on, in the double-slit experiment, this will produce beautiful strips of alternating light and dark.

3.1 Time–Energy Coordinates

3.1.1 *Time–Energy Polar Coordinates*

In our phase plane, not every circle is allowed. To be legitimate, the circle must have energy

$$\frac{m\omega^2}{2}\left(x^2 + \frac{p^2}{m^2\omega^2}\right) = \frac{m\omega^2}{2}|z|^2 = \frac{m\omega^2}{2}\cdot\frac{\bar{z}z + z\bar{z}}{2}$$
$$= \omega\bar{h}\left(n + \frac{1}{2}\right), \quad n = 0, 1, 2, 3, \ldots$$

(Figure 3.1). These circles will help diagonalize the Hamiltonian radially. How? The nth circle will model a more global function: $\bar{z}^n$ (times the ground state). This function will precess (cycle) at a constant frequency:

$$-\omega\left(n + \frac{1}{2}\right).$$

This is the nth energy level (in the complex plane). But it also takes a different face: on the x-axis, it will model a different function

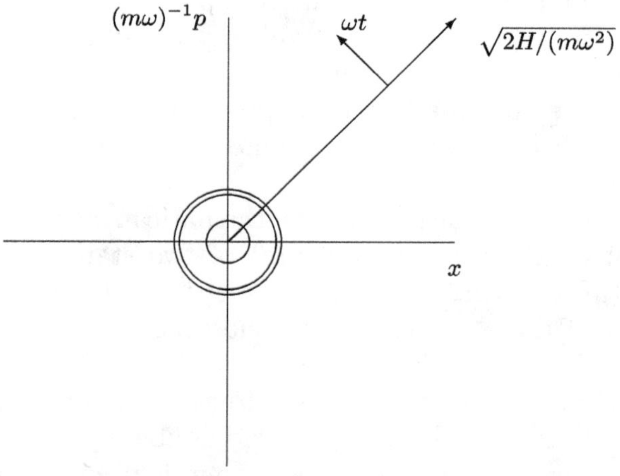

Figure 3.1. Time–energy polar coordinates: energy is radial, and time is angular. Each energy level makes a circle. As time goes by, the ray (issuing from the origin) rotates counterclockwise. The Fourier modes $\bar{z}^n$, on the other hand, precess clockwise (faster and faster, with more and more energy as n grows).

(in x) — an eigenfunction of the Hamiltonian: a polynomial of degree n in x (times the ground state).

How to illustrate this geometrically? On each radial ray (issuing from the origin), let energy grow monotonically outward (as if the Hamiltonian was diagonalized radially). This way, H means energy (unit: mass times velocity squared). This way, our ray makes a new (radial) coordinate:

$$r \equiv |z| = \sqrt{\frac{\bar{z}z + z\bar{z}}{2}} = \sqrt{\frac{2H}{m\omega^2}}$$

(unit: length, as required). Time, on the other hand, is angular (going round and round, like a hand in a clock). It makes our second polar coordinate:

$$\theta \equiv \arg(z) = \omega t.$$

(This is a pure number: no physical unit at all.) Now, how to jump to the next (higher) energy level? Physically, absorb a photon (see exercises in Chapter 17). Mathematically, on the other hand, multiply by $\bar{z}$ (to help "climb" from $\bar{z}^n$ to $\bar{z}^{n+1}$). How to do this geometrically? Look at one particular circle. For this, assume that energy is fixed (although unknown). This means that we already projected onto one particular circle (but don't know which). On it, we have a new uncertainty principle: once you are confined to a specific energy, you can know nothing about time any more! Thus, on the circle, time must be distributed evenly. How to guarantee this? On the circle, the time wave function can only change phase (angle), not magnitude. It must behave like an angular Fourier mode: $\bar{z}^n$. (Compare this to momentum: Section 2.3.3.) This way, energy will be conserved in time, as required. Thanks to this picture, we are now ready to add a quantum of energy.

3.1.2 *How to Add a Quantum of Energy?*

How to add one quantum of energy to $\bar{z}^n$ (and upgrade it to $\bar{z}^{n+1}$)? Apply a new operator:

$$A^+ \equiv \sqrt{\frac{m\omega}{2\hbar}}\left(X - \frac{i}{m\omega}P\right) \sim \sqrt{\frac{m\omega}{2\hbar}}\bar{z}$$

(and renormalize). Later on, we'll also see how A^+ works algebraically and helps advance to the next higher energy level (Section 9.3).

To help normalize, A^+ also divides by the extra factor $\sqrt{2}$ (mirrored in Section 2.5.2). Here, "$\sim$" is in terms of the Fourier transform (and will also mean picking, as discussed in the following). Indeed, what does A^+ do? In other words, how does $X - iP/(m\omega)$ act circularly (angularly)? Well, it picks $\bar{z}$. Indeed, it splits into two polar parts: tangential and radial. How do they act angularly? They act in two different ways: the tangential part is in its diagonal form and the radial part differentiates. For example, what happens at the point $(0, p_0)$ (at the bottom of the circle)? Here, they act rightward, on a wave like $\exp(ixp_0/\bar{h})$:

- $X = \text{diag}(x)$ picks $x = 0$,
- $-iP = -\bar{h}\partial/\partial x$ picks $-ip_0$.

For yet another example, what happens at $(x_0, 0)$ (the rightmost point on the circle)? Here, they act upward, on a wave like $\exp(-ix_0 p/\bar{h})$:

- $P = \text{diag}(p)$ picks $p = 0$,
- $X = i\bar{h}\partial/\partial p$ picks x_0.

In either case, we pick $\bar{z}$, as required. Moreover, how to extend this to the entire circle? Exponentiate:

$$x_0 - ip \to x_0 \exp(-ip).$$

This is relevant even to $p \neq 0$. This is how a Lie algebra (the vertical line) generates a Lie group (the circle). Thus, $\bar{z}^n$ gets multiplied by $\bar{z}$ and upgraded to the next angular Fourier mode, with a higher frequency and more energy: $\bar{z}^{n+1}$ (renormalized).

3.1.3 *How to Make an Eigenfunction?*

Better yet, apply A^+ horizontally (rather than circularly). After all, this is how X and P were designed in the first place: to "look" everywhere, at all x's at the same time (Section 2.1.3). You can even repeat this, time and again: start from the ground state (on the x-axis) and add more and more energy (quantum by quantum), producing higher and higher eigenfunctions of the Hamiltonian (on the x-axis).

Fortunately, both X and P commute with F (Section 2.4.3). Thus, at the same time, you can apply A^+ vertically as well. The complex wave function will remain invariant and keep its original property: the momentum wave function will still be F times the wave function, as required.

How to work horizontally? In the original definition of the function q (Section 1.5.1), interpret x^n as an operator (rather than a mere number). In other words, in our Fourier series (Section 2.5.1), replace $\bar{z}^n$ by the corresponding operator:

$$\bar{z}^n \leftarrow \left(X - \frac{i}{m\omega}P\right)^n .$$

How to apply this horizontally? First, restrict the ground state to its wave function (defined on the real axis):

$$\tilde{g}(x,0) = \exp\left(-\frac{m\omega}{2\hbar}x^2\right)$$

(Figure 9.1). To this, apply A^+, time and again. This will yield the eigenfunctions of the Hamiltonian (one by one), ready to be multiplied by their corresponding coefficients, and superpose the wave function, as required. For the coherent state, for example, these coefficients are already stored in the $\mu/\sqrt{2}$th column in $A^{-*/2}$. By plugging them, we'll obtain the wave function (in terms of x):

$$\tilde{g}(x,\mu) = \exp\left(-\frac{m\omega}{2\hbar}(x-\mu)^2\right).$$

3.1.4 *How to Make an Angular Fourier Mode?*

How to model this geometrically (in general)? For this, work circularly again. As before, the nth circle will model the nth angular Fourier mode:

$$\bar{z}^n = r^n \exp(-in\theta) = r^n \exp(-in\omega t)$$

(times a function that is independent of n). For example, let's use this to obtain the coherent state again (in the entire complex plane). For this, start from the ground state again:

$$\tilde{g}(\bar{z},0) = \exp\left(-\frac{m\omega}{2\hbar}\bar{z}^2\right).$$

To this, apply A^+. How? If horizontally, then you'd get the eigenfunctions of the Hamiltonian again (as functions of x). But here, do something else: on each individual circle, apply A^+ angularly again (time and again). This will introduce the $\bar{z}^n$'s, which precess clockwise (in terms of $\theta = \omega t$): their phase grows clockwise, at frequency $-n\omega$. Once superposed, they can make any polynomial in $\bar{z}$ and even expand the entire Fourier series, as required.

3.2 Energy and Its Distribution

3.2.1 *Static Discrete Wave Function*

Can this change dynamically in time? Only in terms of our old Cartesian coordinates! Indeed, on a (horizontal or vertical) line, a few eigenfunctions of the Hamiltonian can superpose and interfere. Constructively or destructively? Well, this depends on their proportion, which changes all the time (due to their different frequencies). This is physical: it will produce a dynamic wave function, changing all the time. Once squared (and normalized on the entire line), this will be the probability to be at x (on the horizontal line) or to have momentum p (on the vertical line).

This is in our old Cartesian coordinates. In our new polar coordinates, on the other hand, we are more interested in the discrete wave function: a function of n only. In the coherent state, for example, this is the $\mu/\sqrt{2}$th column in $A^{-*/2}$. It has two important jobs:

- It helps expand the Fourier series.
- Once squared (and normalized over all n's), this is the probability to be at the nth energy level. In the coherent state, for example, this is the Poisson distribution.

This is not dynamic but static. After all, in a Fourier series, the coefficients remain the same, all the time.

3.3 Time–Energy Uncertainty

3.3.1 *Time–Energy Uncertainty*

Together, time–energy mirror our old Cartesian coordinates: position–momentum. Indeed, the bigger the momentum, the bigger the wave number: the "kick" to the wave function (that oscillates in x). Likewise, the higher the energy, the higher the frequency: the "kick" to $\bar{z}^n$ (that precesses in time). In the complex wave function, this makes the Fourier series: a sum of products — coefficient (of energy) times precession (in time).

Moreover, time–energy mirror position–momentum in yet another way: they obey their own uncertainty principle. In fact, just like you can't know position and momentum simultaneously, you can't know time and energy simultaneously. How to see this geometrically? The key is area. Inside the nth circle, the area is proportional to both energy and frequency: $n+1/2$. Thus, the area of the annulus between two consecutive circles is constant. How to use this? In Figure 3.1, look at bigger and bigger circles, with more and more energy:

$$r \to \infty \quad \text{and} \quad H \to \infty.$$

On one hand, θ has more and more room to be specified accurately. This way, time gets less and less uncertain. On the other hand, energy grows like r^2: too fast for the r coordinate to catch up. As a result, energy has little room to get specific and deterministic. It grows over the top and can even spread out and leak into neighbor circles (that get too crowded since n grows like $H \gg r$):

$$\left(r^2\right)' = 2r \to_{r\to\infty} \infty$$

or, equivalently,

$$\left(\sqrt{H}\right)' = \frac{1}{2\sqrt{H}} \to_{r\to\infty} 0.$$

Thus, as energy grows, it also gets more and more sensitive, unstable, and uncertain.

What really happens on a big circle? Well, high energy also means high frequency. After all, energy and frequency go hand in hand: they are proportional to one another (Planck's law). Thus, as n grows, precession gets faster and faster. In theory, this is the whole story. In practice, on the other hand, there is more. High energy can get quite unstable, and even leak, and radiate light: an outgoing photon, carrying the extra energy. In our geometrical picture, this looks like falling down (to a smaller circle, with less energy).

The higher the initial energy, the shorter its lifetime. Low energy, on the other hand, is more stable and attractive. This will give us a practical guideline, useful in AI too: filter out high energy and focus on low energy only. This way, even with noisy data, full of errors, your model will still make sense, and your predictions will still be within reason. Later on, we'll see a few interesting examples (from image processing). Before going into this, let's look at what we did so far in quantum mechanics and draw some interesting conclusions about dynamics and entanglement.

3.4 Dynamics in Time–Energy

3.4.1 *Heisenberg or Schrodinger Picture?*

The time–energy uncertainty principle makes the world tick. It even supports the most elementary state: vacuum. In fact, vacuum is not quite empty but full of virtual particles that exist and don't exist at the same time. Thanks to the time–energy uncertainty principle, for a very short while, a little energy could even be "borrowed" from the future and help produce a new (virtual) particle, ready to disappear instantaneously. This follows from quantum field theory.

This is why our complex wave function is evaluated at $\bar{z}$, not z. Initially, X acts rightward, on the initial wave function. As time goes by, z precesses (rotates) counterclockwise. For us, this is now the new present: the new initial time. In it, we now have new operators, along oblique rays: X acts now up-right (rather than rightward), and P up-left (rather than upward). Wave functions, on the other hand, never change. This is the Heisenberg picture.

Better yet, look at the Schrodinger picture. In it, things are the other way around: operators never change. Only wave functions change. Initially, the wave function is defined on the real axis. This is how it tells us about the initial position. But now, don't forget: we are already at our new initial time (the new present). Now, we have a new complex wave function. Does it agree with the old one? Fortunately, it is evaluated at $\bar{z}$, which already precessed clockwise: back in time, into the past. In terms of our new (shifted) time-scale, this is just a new name for the old initial point, which is now in the past. At this point, the new complex wave function will be designed to agree with the old one. In what sense? Well, the old wave function told us about the initial position. But the new one can only tell us about energy: after all, in our new coordinates, the new $\bar{z}$ is no longer on the real axis but on an oblique ray. In this direction, it can't tell us about position but only about energy. Still, in the sense defined in the following, the new and old functions will still agree and be invariant. This will help model spacetime circularly: a Riemann surface.

This is kind of relativity: a new time-scale, starting from a new present. Let's extend this to energy as well. This will help model anti-matter too.

3.4.2 *Toward Anti-Matter*

This kind of relativity applies to not only time but also energy. To see this, let's look at things in a wider context. In quantum mechanics, time doesn't flow at all. It is a random variable, just like energy. Thus, the same kind of relativity (used to have a new present) is relevant to energy too, with two changes:

- Instead of numbers, use operators.
- Instead of advancing in time, advance to the next (higher) energy level. For this, apply A^+ angularly. In our phase plane (in terms of the Fourier transform), this is

$$
\begin{aligned}
A^+ &\sim \sqrt{\frac{m\omega}{2\bar{h}}}\bar{z} \\
&= \sqrt{\frac{m\omega}{2\bar{h}}} \cdot \frac{\bar{z}z + z\bar{z}}{2z} \\
&= \sqrt{\frac{m\omega}{2\bar{h}}} \cdot \frac{2H}{m\omega^2} z^{-1} \\
&= \sqrt{\frac{m\omega}{2\bar{h}}} \cdot \frac{2(n+1/2)\omega\bar{h}}{m\omega^2} z^{-1} \\
&= \left(n + \frac{1}{2}\right)\sqrt{\frac{2\bar{h}}{m\omega}} z^{-1}
\end{aligned}
$$

(Section 3.1.2).

Geometrically, on each individual circle, A^+ mirrors $\bar{z}$. It looks like an arrow, issuing from the origin and pointing at $\bar{z}$ (like a hand in a clock). Algebraically, on the other hand, this arrow mirrors 1, which spans $\mathbb{Z}$: the additive group of integer numbers (just like $\bar{z}$ spans the entire circle). As a bonus, the above formula helps model anti-matter too: inverse powers of z, back in time.

Likewise, in quantum field theory, we'll work with operators only, not numbers at all. The important object will be the operator (that creates a new photon or so), not the initial state on which it acts. For this reason, even the wave function will return an operator, not a number any more. Even the phase plane will be made of operators, not numbers any more. In this context, our (complex) PDE should use a Fréchet derivative (Section 2.5.6).

3.4.3 *Projection on a Curve*

To model anti-matter, A^+ acted on each individual circle (although we don't know which). How to do this in general? In our phase plane, consider some physical quantity. For it, pick some (hypothetical) value. Once this is picked, what happens to the original complex wave function? In a sense, it gets projected (or confined, or restricted) to some (1-D) curve. How to use the restricted function? Do three things:

- take its absolute value,
- square it up, and
- normalize (to have integral 1 on the entire curve).

This is the probability density for some other (dual) physical quantity, distributed along the entire curve.

We already saw a few examples. To project on a horizontal line, pick a value for momentum. This would reproduce the good old wave function back again (in terms of position: x). To project on a vertical line (in terms of a hyperbolic metric), on the other hand, pick an x. This would reproduce the momentum wave function back again (in terms of p). To project on a circle, pick an energy level. And, to project on a ray (issuing from the origin), pick a time.

3.5 Dynamics in the Coherent State

3.5.1 *The Coherent State and Its Dynamics*

For example, in the coherent state, look at the complex Gaussian. To have the wave function (in terms of position), restrict to the real axis:

$$\tilde{g}(x,\mu) = \exp\left(-\frac{m\omega}{2\bar{h}}(x-\mu)^2\right).$$

Now, let time go by, and look at a new (fixed) time: $t > 0$. For us, this is now the new present: the new initial time. So, our new x-axis is no longer horizontal: it already rotated (counterclockwise) by angle ωt (Figure 3.2). Perpendicular to it, we now have a new imaginary axis. Use these new axes to span a new complex plane. In it, we can still look at the Schrodinger picture, which still keeps the same operators (with the same meaning as before): X still means position (along the

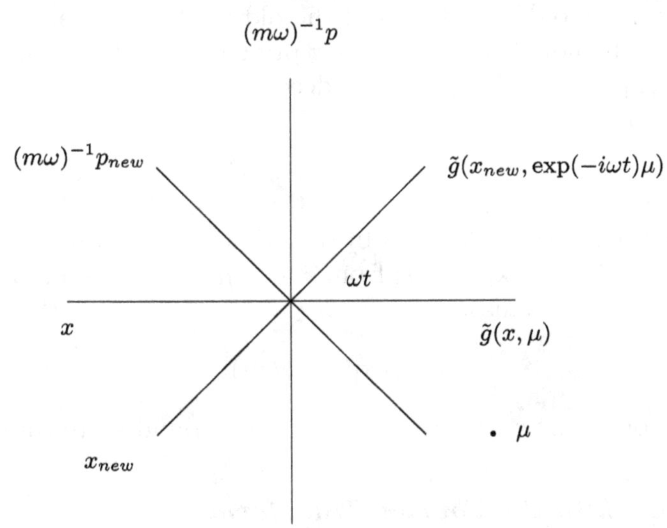

Figure 3.2. Dynamics: the Schrodinger picture. At the new time $t > 0$ (the new present), the new position axis is now oblique (pointing up-right). On it, the new wave function (in terms of x_{new} rather than x) is now obtained from a new complex Gaussian (in which μ already precessed by angle ωt clockwise).

new real axis) and P momentum (along the new imaginary axis). But now, in our new (rotated) plane, we have dynamics: a new complex Gaussian. Project it on your new x-axis and obtain the new wave function (in terms of the new position x):

$$\tilde{g}(x, \exp(-i\omega t)\mu).$$

This is the dual face of nature: the particle is also a wave, flowing and propagating (and oscillating), all the time. To see this, just look at μ, and see how it changes and oscillates, all the time. This models well the classical harmonic oscillator (from classical mechanics). But here, in quantum mechanics, this is a nondeterministic flow: forward in time, and back in time, at the same time.

Still, this is a bit geometrical. How to see this more analytically? Well, in terms of the old position x, the new present is at

$$z = \exp(i\omega t)x.$$

This is still the old z, in terms of the old x. How to transform both z and x to the new ones (at the new present)? For this, multiply the old z by $\exp(-i\omega t)$. This would indeed yield

$$z = x$$

(the new ones, at the new present). Or, if you like, multiply the old $\bar{z}$ by $\exp(i\omega t)$. How to do this? In the original Fourier series (Section 2.5.1), throw factor $\exp(-i\omega t)$ from $\bar{z}$ onto μ. This would indeed yield the new complex Gaussian:

$$\tilde{g}(x, \exp(-i\omega t)\mu)$$

(in terms of the new x), provided that the ground state rotated too.

3.5.2 *Back to the Fourier Transform*

This is dynamics for you: the wave function changed and evolved; it is obtained now from a new complex Gaussian, in which μ precessed clockwise. (If you like, this is just a new name for the old μ, using the new coordinates.) Thanks to this, at our new present (marked by the new position axis), we now have a real (physical) dynamics: a really new wave function (not only mathematically but also physically). This is thanks to interference: different components in the Fourier series precess at a different frequency, superpose at a new proportion, and project on the new position axis, to form a new wave function.

Is this familiar? We already saw this, in the Fourier transform (in the harmonic oscillator: Section 2.5.6). Indeed, look at the original wave function (in terms of the original position: x). From it, how to obtain the momentum wave function (in terms of p)? No need to apply the Fourier transform explicitly. Instead, look at the dynamics: the bigger the momentum at $x = 0$, the farther will the particle be pushed rightward, and the bigger will x be after a quarter of a period:

$$\omega t = \frac{\pi}{2}.$$

By this time, what will happen? In the Heisenberg picture, the operators will change: P into X, and X into $-P$. Better yet, look at the Schrodinger picture. By this new time, the entire complex plane will rotate (by angle $\pi/2$) counterclockwise: the horizontal x-axis will get vertical (with X acting upward), and the imaginary axis will get horizontal (with P acting leftward, both in their diagonal form). These new coordinates will mark our new present: the new initial time. They will also span our new (rotated) plane. In it, define a new complex Gaussian from scratch:

$$\tilde{g}(z, -i\mu),$$

where z is new. Confine it to its real part, and obtain the wave function (as it looks like after a quarter of a period), as required. Better yet, transform to the old z back again. How to do this? How to recover the old present, which is now in the past? Rewind, back in time:

$$\tilde{g}(-iz, -i\mu).$$

Here, things are different: $-iz$ is new, and z is old. Confine z to its imaginary part, and obtain the momentum wave function, as required. This is what we already did in Figure 2.3. (Pick $c = 0$ there.)

3.6 Squeezed State

3.6.1 *Squeezed State*

So far, we only looked at a constant frequency: ω. Assume now that the frequency suddenly jumps from ω to

$$\tau > \omega.$$

As a result, our coherent state gets squeezed, with only a small (horizontal) variation around $\bar{\mu}$. Still, how does it look like in terms of the old energy levels? Well, it can still be expanded as a linear combination of the columns of $A^{-1/2}$ (although rescaled to have a new ground state, corresponding to τ rather than ω) but with new coefficients:

$$\exp\left(-\frac{m\tau}{2\hbar}\mu^2\right)\frac{\left(\sqrt{\frac{m\tau^2}{2\omega\hbar}}\mu\right)^n}{\sqrt{n!}}.$$

Once squared (and normalized over all n's), these give us a new Poisson distribution:

$$\exp\left(-\frac{m\tau^2}{2\omega\hbar}|\mu|^2\right)\frac{\left(\frac{m\tau^2}{2\omega\hbar}|\mu|^2\right)^n}{n!}$$

($0 \leq n < \infty$). For a large n, this is much bigger than before, especially when

$$\tau \gg \omega.$$

Thus, in its new (squeezed) state, the particle is now much more likely to be highly energetic. How to use this?

3.6.2 *Toward The Big Bang and Inflation*

Right after the big bang, our infant spacetime expanded and inflated in size, exponentially fast: even faster than light. Our original particle, on the other hand, had no time to follow and stayed at the middle. From the far edge of the universe, it looked very small (due to length contraction), and its time very slow (due to time dilation). Its frequency (which behaves like 1/time) suddenly jumped. Its state squeezed and became highly likely to be very energetic. New energy also means new mater, as in Einstein's formula:

$$E = mc^2$$

(at least in the self system of the particle itself). These new energy and matter radiated and spread out into the entire universe and gradually fused into more particles, atoms, and even molecules. Later on, this new matter cooled down, and formed stars and even galaxies. This is how something came from nothing: a new universe was born [15]. (For exercises, see Chapter 7.)

Chapter 4

Toward Entanglement and The EPR Paradox

What is the main principle behind quantum mechanics? Interference! Is this mathematical or physical? It is more mathematical! In this chapter, we highlight the mathematical structure that designs interference: duality and indeed (anti-)symmetry. We illustrate this in a lot of examples: electromagnetics, spin, anti-matter, and more. This is where nature submits to mathematics and takes its structure from group theory.

Spin leads us to entanglement: the basis for quantum computing. But entanglement is strange: it leads to the EPR paradox. To answer it, we'll look at our own (implicit) logic that hides behind entanglement. Is entanglement real and physical? I don't know. You tell me!

4.1 Toward the Path Integral

4.1.1 *Reversibility and the Path Integral*

So far, we only looked at the 1-D case: $d = 1$ and $x, p \in \mathbb{R}$. Consider now a more realistic state: a new wave function, defined on a 2-D (or even 3-D) surface in spacetime. What is its dynamics? How does it change in time? How does it evolve and develop? How does it convert into a new wave function, defined on a new surface in spacetime? To see this, let's focus on one particular point on the original surface. This is the initial event. From it, many paths issue, leading to the new surface. One of them is the shortest, arriving first (in terms of its own proper time, measured in a tiny clock, traveling along the path). In classical physics, only this path survives. All the others, on the other hand, are gone. With them, a lot of information gets lost. This is the second law of thermodynamics: entropy grows, with no return.

In quantum mechanics, on the other hand, things are much more reversible. All paths survive and could materialize (at least in theory): short ones at a high probability and long ones at a low probability. On the new surface in spacetime, they interfere and superpose a new wave function. This is the path integral.

Each path precesses in its own rate: its private frequency. To obtain this frequency, integrate the Lagrangian (which is closely related to the Hamiltonian) along the path (with respect to its proper time). This is its own (private) frequency: it will tell us how it should interfere with other paths and superpose at a new point on the new surface in spacetime.

This interference (constructive or destructive) will help produce the new wave function, which will tell us the probability to arrive at this new point. But this remains theoretical: we don't get to see this (beautiful) wave pattern, until we actually measure (on a screen, placed there).

So long as we don't measure, the process is still reversible. After all, we can still rewind the movie and look at it the other way around: back in (proper) time. On each individual path, precess in the opposite direction, back to the original wave function, at the initial point, on the original surface in spacetime. This is also how a Feynman diagram works: an electron absorbs a photon and jumps to a higher energy level. Or, if you like, look at things the other way around (in terms of anti-matter): a positron (a missing electron) spits a photon and falls down (to a lower energy level), back in time [32].

4.2 The Double-Slit Experiment

4.2.1 *The Measurement Paradox*

A good example is the double-slit experiment (Figure 4.1). To see this, assume that time flows rightward (horizontally, with the particle), and that space is vertical. In this spacetime, there are only

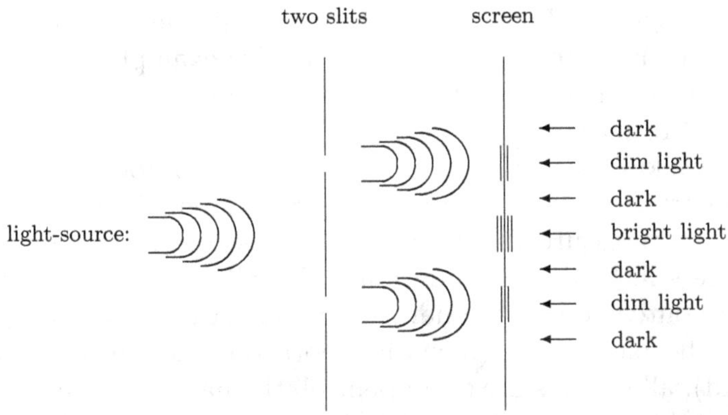

Figure 4.1. At each slit, the light diffracts and restarts like a new wave rightward. On the screen behind, these two waves interfere and make alternating strips of light and dark.

two paths: one through the upper slit and the other through the lower slit. On the screen, on the far right, we get to see how they interfere and make new (beautiful) strips of alternating light and dark. This is because we measure: the screen is our measuring device.

In the path integral (Section 4.1.1), things are still reversible. But, upon measuring, things change dramatically. Many paths interfere destructively and disappear. After all, our measurement already proved them wrong. Only a few survive: the shortest ones. After all, our measurement already proved one of them right.

In our experiment, only two paths get to survive: either through the upper slit or through the lower. After all, both are as short (and as likely). On the screen behind, they interfere (or superpose, or sum up). After all, our measurement already proved them right.

Still, what about reversibility? Didn't we say that all paths must survive, to let us march on them (back in time), if we like? And what about entropy? In classical physics, it grows (information gets lost). But here, shouldn't entropy remain the same, with no loss of information at all?

4.2.2 *Measurement Is Entanglement*

Well, the above picture describes one (lonely) particle only. But, once measured, the particle is no longer alone: it is now connected (entangled) to our measuring equipment. Together, they make one big united system. In it, interference is real: measurable and even reversible (thanks to the wave pattern). For example, in a Feynman experiment, the higher the precision, the more you know about more paths used to measure.

Once we look, we let our mind join the system too and become an integral part of it, producing a yet bigger system. This is how we get to "see" or "measure": our mind entangles too. In this big system, things are still benign and reversible. Why? Because we didn't pick a favorite outcome yet. We never really "measured" but only entangled! In the big system (including both our measuring device and our mind), all options are still open. Mathematically, the big system is more stable: in it, the wave function still keeps its original norm. This way, in theory, we can still rewind (back in time) and forget.

Indeed, there is a way to spoil the double-slit experiment altogether. Before the photon reaches the screen behind, detect which slit was picked: upper or lower? Once the answer is recorded, there will be no interference any more! Why? Because your system is now even bigger: it contains now the detector too. To help the wave function keep its original norm, the detector splits it into two separate parts. This way, our two paths have orthogonal wave functions that "live" on two parallel hyperplanes and never correlate or interfere at all any more. Later on, we'll see how entropy splits between two parallel universes, allowing reversibility.

4.3 Duality and (Anti-)Symmetry

4.3.1 *Dual (Anti-Symmetric) Nature*

Why is math so good at modeling nature? Because math is inherently dual. Indeed, look at the unit circle. On it, you can move counterclockwise, then clockwise, and get back to your starting point. This is the key to complex numbers, leading to interference. Indeed, in the complex plane, you have yet another kind of (anti-)symmetry: take the complex conjugate, time and again, and get back to your starting point. Actually, this comes from elementary algebra:

$$(-1)(-1) = 1.$$

This is an elementary switch permutation: apply it twice and get the same picture back again.

Permutation is the key to the determinant, leading to vector product: an elementary Lie algebra, good to model more advanced physics, including angular momentum and electromagnetics. The magnetic dipole is indeed dual again: interchange north and south, time and again, and get the original field back again.

In quantum mechanics, angular momentum is discrete, not continuous any more. In its simplest form, it is dual again and has just two states. This is enough to explain two interesting phenomena: spin and polarization (Chapters 10 and 11). This elementary kind of duality is good enough to model a very simple system: just one electron (or photon). To model a few electrons (in their energy level in the atom), we already need more advanced duality: Slater's determinant leads to the Hartree–Fock system.

4.3.2 *Toward Anti-Matter*

And what about matter in general? Well, it is dual too. Indeed, on top of spin, Dirac introduced yet another kind of duality [32]: a wave function could cycle in two ways, either clockwise (left-hand) or counterclockwise (right-hand). These are interchangeable again (with an extra minus sign). This leads to anti-matter: underneath the zero-point (ground state), a missing electron (of negative energy) is also called positron (of positive charge and energy). These are interchangeable again (with an extra minus sign). This way quantum mechanics lives in peace with special relativity and its dual (hyperbolic) spacetime. After all, why not go the other way around: back in time?

This leads to duality in Feynman diagrams too. For example, vacuum is not quite empty: it contains a lot of virtual particles (that exist and don't exist at the same time). After all, time could flow either way: either forward or backward (which are both as likely). This is also why two electrons repel one another: they exchange a virtual phonon (either forward in time or back in time, in terms of anti-photon). Thanks to the (time–energy) uncertainty principle, the required energy is "borrowed" and paid back again, too fast to violate any conservation law. In reality, only one thing can really "exist": a tiny loop in spacetime, creating and annihilating back again, as fast as ever.

Is there an anti-universe, made of anti-matter? Perhaps. In it, time flows the other way around: backward. In it, there are anti-people, made of anti-matter. But they think that they are ordinary people and that we are anti-people, and they are probably right!

4.3.3 *Group Theory and the Quantum Jump*

Actually, this is an elementary symmetry group: mirror, and mirror again, and get the same picture back again. This is also the binary group Z_2, containing just two numbers: zero and one. Indeed, start from zero and add one twice and get zero back again. Or, if you like, this is also the Boolean group ± 1. Indeed, start from 1 and multiply by -1, time and again, and get 1 back again.

This is indeed spin: either spin-up (counterclockwise) or spin-down (clockwise). There is no other possibility in between. This is indeed a quantum jump: from angle $-\pi/2$ (spin-down) to angle $\pi/2$ (spin-up), with no intermediate angle in between. Why? Thanks to the uncertainty principle, position and momentum don't commute but have a nonzero (constant) commutator. In the harmonic oscillator, this leads to a jump from one energy level to the next one, with nothing in between. This is how Einstein explained the photoelectric effect: only a high-frequency photon (violet, or at least blue) has enough energy to kick an electron to a higher energy level (or even off the atom altogether).

Likewise, in angular momentum, not every angle is allowed but only a few discrete angles between $-\pi/2$ and $\pi/2$. In the extreme case of spin, the jump is as big as π: no legitimate angle in between at all. All this is only thanks to the uncertainty principle, which is not only due to our inaccurate observation but also real physics: nature is random and nondeterministic at its very heart! This is why both energy and angular momentum are only discrete, not continuous any more.

In the harmonic oscillator, what are the energy levels? So far, they mirror $\mathbb{N}$ (the natural numbers). But this is not a group. To make a legitimate group, don't forget the negative numbers. This is indeed what Dirac did for us: underneath the zero level, he also introduced many more negative levels. An electron can now occupy not only a positive but also a negative energy level (at least in theory). There, a missing electron (of negative energy) is also called a positron (of positive charge and energy). This is how Dirac extended $\mathbb{N}$ into $\mathbb{Z}$: the (additive) group of all integer numbers, positive and negative alike.

Still, our real world is full of electrical currents, made of electrons, not positrons. Why are positrons so rare? Because they are too unstable. Mathematically, -1 is soon multiplied by -1 again, and $\mathbb{Z}$ becomes $\mathbb{N}$ back again.

4.3.4 *Gravity vs. Quantum Mechanics*

Gravity, on the other hand, is inherently different. In gravity, we are only confined to real numbers, not complex. In particular, $\partial/\partial x$ helps design not momentum but curvature (in a new four-dimensional differentiable manifold: spacetime). The same holds for the other coordinates as well.

This is why gravity has no duality or anti-symmetry or any other group structure. To design spacetime geometrically, these are not really needed (at least until a new complex field is defined on it, as in quantum mechanics or electromagnetics). Without complex numbers, gravity can exhibit no interference or uncertainty at all. It is quite deterministic and classical!

In spacetime, how does gravity design curvature? It uses differential geometry: throw dx and dt to the denominator and obtain $\partial/\partial x$ and $\partial/\partial t$ (which help draw curvature). What are these physically? Look at their physical units: 1/length and 1/time, which must stand for wave number and frequency. In quantum mechanics, these are (proportional to) momentum–energy (thanks to the Fourier transform, in space or time). Indeed, just multiply by $\bar{h}$, and you'll get the correct momentum–energy units. But here, in general relativity, momentum is no longer perpendicular to position (as in our phase plane). It is no longer imaginary but real (just like x). This is because $i = \sqrt{-1}$ is now missing. Where is it hiding? Don't worry, it is still there. It was only doubled: $i^2 = -1$. This is how momentum–energy curve spacetime and give it its new metric. This is why Einstein's equations are real (not complex) and hyperbolic (not elliptic): time picks a minus sign. (Is this familiar? This is duality again, in the back door.) Indeed, once linearized (around a flat metric), Einstein's equations reduce to a few (hyperbolic) wave equations. This is how duality can resurface. Until then, it remains implicit: no imaginary numbers, and no room for uncertainty or quanta at all.

4.4 Riemann Sphere and Surface

4.4.1 *Riemann Sphere: Phase Plane*

What can we learn from general relativity and gravity? Curvature! So far, our phase plane was flat: the complex plane. How to curve it? How to turn it into a sphere? Look at the origin as the south pole. Then, foil inward and map from infinity to the north pole. This is the Riemann sphere.

Topologically, it is still the same as the original complex plane. In other words, they are homeomorphic: each open set (without its boundary) is mapped to an open set. Algebraically, on the other hand, the Riemann sphere is slightly better: it makes a new Lie group, with a nonzero commutator. Indeed, if you are on the northern hemisphere, then walking north and then east is shorter than walking east and then north. Likewise, if you are on the southern hemisphere, then walking south and then east is shorter than walking east and then south. This is thanks to the new (positive) curvature. Still, keep the same arithmetic operations as before: to multiply two points, go back to the original complex plane (with its new metric).

Why is this better? Because this mirrors quantum mechanics better. Indeed, quantum mechanics uses a Lie algebra, containing operators that may not commute. This way, the corresponding physical quantities can't be measured simultaneously. This is the uncertainty principle.

For example, look at time–energy. How to mirror them geometrically? Look at Figure 3.1 again. How to map it to the Riemann sphere? Well, each circle is mapped to a latitude, and each ray to a longitude. This is the uncertainty principle again, in its geometrical face. Indeed, latitudes don't commute with longitudes: you must specify in what direction you want to walk first!

4.4.2 *Riemann Surface and Spacetime*

The path integral works in spacetime. But spacetime is not necessarily Cartesian. In the harmonic oscillator, it is a Riemann surface. This is a sequence of infinitely many spheres, enveloping one another. They only share one longitude: say, the 0°-longitude (at Greenwich). This way, time is still circular but no longer periodic. It grows monotonically and "climbs" from sphere to sphere, smoothly, through their joint longitude. And what is the dynamics? Well, we already saw it (Section 3.5.1). This is a kind of path integral, with two changes:

- polar coordinates (rather than Cartesian) and
- uncertain position (in both the beginning and the end).

This is indeed how a wave propagates.

4.5 Parallel Universes: Can Entropy Split?

4.5.1 *The EPR Paradox*

Imagine two particles, created in the same physical process. They only differ in one aspect: they fly in opposite directions. So, they must have the same spin: either both up or both down. Why? Because they are entangled (at least in terms of spin).

Now, let them fly for a long time (away from one another). Then, measure the spin of one of them. Immediately, you get to know the spin of both. (After all, it is the same.) How come? How could information arrive to you faster than light? This is the EPR paradox (Einstein–Podolsky–Rosen).

This is a good question, which deserves many answers. (Why so many? Because none is good enough.)

1. Physically, the particles were never entangled. Each was only entangled to the equipment that produced it in the first place. We only used our own logic (which is indeed faster than light) to assume that the equipment always works in the same way. We only called them "entangled" to point out their correlation.
2. Once the particles part from one another, their joint (global) wave function must change considerably. Their spin, on the other hand, remains the same. Why? Thanks to conservation laws. But this is our logic again. Physically, we don't know this, until the good news travels all the way back to us: "Don't worry: spin didn't flip!

It is still the same, independent of time or space!" But this takes time: it is not faster than light at all.
3. Physically, the particles may keep in touch even over a long distance. For instance, if they have charge, then they must exchange a virtual photon all the time, to carry their electrostatic force.
4. Besides, on their way away from one another, they get hit by many virtual particles (coming from vacuum energy) and get disentangled long before measurement. (This is probably why nothing is faster than light: too many virtual particles stand in the way!)
5. A particle is not quite local: it must also carry some global information. Indeed, to locate it, we must look. But where? Everywhere! But this is global again. Indeed, once we know it is here, we also know it is nowhere else. In quantum field theory, the particle has its own (global) field, which "lives" throughout the entire universe.
6. Before measurement, entropy splits among multiple (parallel) universes. Upon measurement, only one universe survives (for us). This is new information, but it comes at a price: thanks to the uncertainty principle, we also lose information about other properties that change forever, with no return. Moreover, the measurement requires an experiment, which is a long process in its own right. It generates heat, which increases entropy back again. In our case, for example, we start from two (hypothetical) universes. In one, both particles spin up. In the other, both spin down. Where are we? To find out, carry out an experiment: measure the spin of one particle.

The EPR paradox has yet another version: for one particle, measure spin-up, either 0 (up) or 1 (down). For the other particle, let someone else measure spin-right and send the result back to you. This way, for the first particle, you get to know both spin-up and spin-right at the same time, in violation of the uncertainty principle! Something strange is going on: spin-right is not quite a physical property but only a random variable, born with an experiment or an observation.

4.5.2 *Is the Moon Still There?*

Einstein had doubts about quantum mechanics. He asked, when I don't look at the moon, is it gone? Not quite. It is only disentangled from you. Want to see it? Just look! This way, you join the same system. Photons can now come from the moon to your eyes, and let you see the light. They also carry some more good news: don't worry, the moon is still there!

On their way, they get hit by a lot of virtual particles (coming from vacuum energy). Fortunately, they can still go through and arrive. But, if you turn your back, then no photon can arrive to your eyes any more. Want to know reality, as it really is? Look and entangle!

4.5.3 *Kant and the Bigger System*

How to discover the true laws of nature? For this, we need physical experiments and observations. This is how Galilei pioneered true science. On top of this, Newton introduced calculus to help model gravity and astronomy. Thanks to this historical background, Kant was ready to introduce his own philosophy.

Kant realized that science must be independent of our state of mind or point of view. The absolute truth follows not from our sensory input but from objective experiments and data and their analysis. Together, these will form a complete mathematical theory, based on pure logic and rational reasoning.

This fits well with Einstein's theory. After all, both special and general relativity are invariant: switch to new coordinates, and you'll still get the same nature, with the same dynamics, equations, and laws.

In quantum mechanics, on the other hand, things are not so objective any more. In the double-slit experiment, for example, to see interference on the screen behind, we mustn't know which slit was used. This is what puzzled Einstein so much: how come? How could science get back to the ancient days and become subjective back again? And what about all the lovely progress made in (special and general) relativity?

Don't worry: science is still objective, not subjective. Interference gets spoiled not because you looked at the slits but because you measured and recorded the results (even if you never read them). Why? Because the system gets bigger: it contains now all the experimental equipment, including the measuring device, used to find out which slit was used (Figure 4.1).

The new (bigger) system has more degrees of freedom, orthogonal to one another. Even when superposed, they remain separate and don't interfere any more. This is indeed quantum mechanics for you: careful science, supplemented with math — group and distribution theory. (For exercises, see Chapter 8.)

Chapter 5

Application: Image Processing and Energy

From quantum mechanics, we can now learn fundamental principle, useful in AI too: minimum energy. Indeed, as we saw above, low energy is less uncertain and more stable and attractive. Better disregard high energy altogether and focus on low only. This is relevant in machine learning as well. For this, we must first discretize: move to a finite discrete system.

In quantum mechanics, we have an important analytic tool: the Fourier transform. Indeed, we already saw it in its dual face. Physically, it is interference. Geometrically, on the other hand, it rotates the entire complex plane. Either way, it is continuous. This is how it acts on the original wave function and produces the momentum wave function. From it, we can then deduce how momentum is distributed.

In image processing, on the other hand, we work on a digital image. For this, we must discretize. How? Sample the Fourier modes on a uniform grid and obtain their discrete version. This mirrors the energy levels, used in quantum mechanics.

The discrete Fourier modes will help in two major applications: pattern recognition and compression. In both, we'll use the same principle again: minimum energy, to have maximum stability. Later on, we'll also see an efficient algorithm: FFT (fast Fourier transform).

5.1 Discrete Fourier Transform

5.1.1 *Discrete Fourier Transform*

How is this related to AI? To see this, we need some Fourier analysis. So far, we used the continuous Fourier transform: an integral over $x \in \mathbb{R}$. This way, for a fixed momentum p, the wave (Fourier mode) is obtained from the inverse Fourier transform. This is a function of x, defined on the horizontal line. How does it look like? It oscillates (in x):

$$\exp(ikx) = \exp\left(ix\frac{p}{\hbar}\right).$$

This is the kth Fourier mode, of wave number

$$k = \frac{p}{\hbar}.$$

This also works with time–energy: on the nth energy level, the (angular) Fourier mode precesses like $\bar{z}^n$. If you like, this is also a column in $A^{-1/2}$: a continuous function of z: z^n (applied to $\bar{z}$, rather than z). How to discretize it? Sample at a few discrete z's: roots of unity (Figure 5.1). This will produce the new Fourier matrix:

$$F \equiv F(N).$$

But this is not yet unitary. How to make it unitary? Normalize it:

$$F \leftarrow \frac{1}{\sqrt{N}}F.$$

This way, F gets unitary:

$$F^* = F^{-1}.$$

This is in 1-D: $d = 1$ and $x_n \in \mathbb{R}$. In 2-D, on the other hand, take the tensor product.

5.1.2 *How to Diagonalize the Laplacian?*

Furthermore, F diagonalizes our good old A: the discrete Laplacian. More precisely, to avoid singularity, A also contains another term: the identity matrix (as in implicit time-marching):

$$A \doteq -\triangle + I.$$

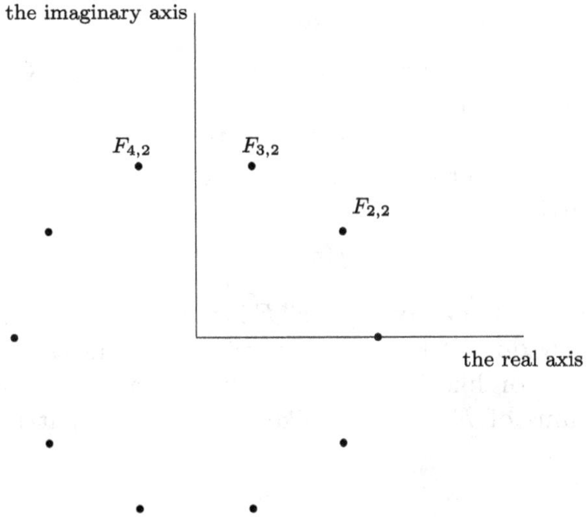

Figure 5.1. The first Fourier mode is the second column in F (before being normalized to get unitary). At the top, we have the Nth root of unity: $F_{2,2}$. Just below it, we have its square: $F_{3,2} = F_{2,2}^2$. Just below it, we have $F_{4,2} = F_{2,2}^3$, and so on. The next column oscillates twice as fast, the next three times as fast, and so on. (In the end, don't forget to divide by $\sqrt{N}$, to get unitary.)

To discretize, A uses second-order finite differences on a uniform grid, with periodic boundary conditions. This way, F indeed diagonalizes A:

$$A = F^*DF,$$

where D is diagonal:

$$D = \text{diag}(d_1, d_2, d_3, \ldots, d_N).$$

On its main diagonal, D contains the (positive) eigenvalues of A (whose eigenvectors are the columns of F^*).

How to use this? In machine learning (Section 1.5.1), we solved for infinitely many w_n's: the coefficients of the columns in $A^{-1/2}$. (Don't confuse them with the roots of unity in Figure 5.1.) But here, things are even simpler: we have only N w_n's, and $A^{-1/2}$ is now a square $N \times N$ matrix:

$$A^{-1/2} \equiv F^*D^{-1/2}.$$

(This justifies the name of $A^{-1/2}$.) Better yet, let the w_n's be the coefficients of the columns in F^*, which are more general, and don't depend on A at all.

5.1.3 *Spectral Approximation*

Better yet, make F^* rectangular. For this, pick only $K < N$ leading columns, and drop the rest. This way, F^* becomes an $N \times K$ matrix, whose columns are smooth Fourier modes. In F, on the other hand, pick only the K upper rows, and drop the rest. This way, F becomes a $K \times N$ matrix. By now, we still have

$$FF^* = I$$

(the $K \times K$ identity matrix), but F^*F is not quite the identity matrix any more. How does it look like? It splits: it behaves either like the identity matrix or like the zero matrix. On smooth Fourier modes (the K columns of F^*), it acts like the identity matrix and leaves them unchanged:

$$(F^*F)\,F^* = F^*\,(FF^*) = F^*I = F^*.$$

More oscillatory Fourier modes (the $N - K$ columns that were dropped from F^*), on the other hand, are in its null-space and are mapped to zero. Next, cut D short:

$$D \equiv \text{diag}(d_1, d_2, d_3, \ldots, d_K)$$

(a new $K \times K$ diagonal matrix, containing K small eigenvalues only). Finally, look at w (the vector of unknowns), and cut it short as well:

$$w \equiv (w_1, w_2, w_3, \ldots, w_K)^t$$

(a new K-dimensional vector of unknowns). This way, the new w_n's ($1 \le n \le K$) will serve as coefficients for the K columns in F^*. Thanks to the above, we still have a good spectral approximation of A:

$$A \doteq F^*DF.$$

Indeed, both sides agree on their joint eigenvectors — the smooth Fourier modes (the K columns of F^*):

$$(F^*DF)\,F^* = F^*D\,(FF^*) = F^*DI = F^*D = AF^*.$$

And what about the inverse matrix? It is approximated well as well (in the same way):

$$A^{-1} \doteq F^*D^{-1}F.$$

In what sense? Again, both sides have K joint eigenvectors — the smooth Fourier modes (the columns of F^*):

$$\left(F^*D^{-1}F\right)F^* = F^*D^{-1}\left(FF^*\right) = F^*D^{-1}I = F^*D^{-1} = A^{-1}F^*.$$

How to use this?

5.2 Principle of Minimum Energy

5.2.1 *Energy Minimization*

We are now ready to minimize our new energy:

$$\frac{1}{2}\left(F^*w, AF^*w\right) = \frac{1}{2}\left(F^*w, F^*Dw\right) = \frac{1}{2}(w, Dw),$$

subject to

$$F^*w = b.$$

But these are N constraints in just K unknowns. On one hand, this is good: more stability, and less noise or overfitting. On the other hand, this is overdetermined and can be solved only in terms of least squares. For this, multiply both sides by F (from the left):

$$w = Iw = \left(FF^*\right)w = F\left(F^*w\right) = Fb.$$

These will be our K new constraints.

5.2.2 *Relative Importance*

By now, are we on the right track? We sure are. After all, our new energy $(w, Dw)/2$ uses only smooth Fourier modes, avoiding

- high energy,
- high frequency,
- high variation, and
- high fluctuations.

Furthermore, we only use $K < N$ degrees of freedom, which is good for stability and avoids noise and overfitting.

By now, our system is no longer overdetermined. Indeed, we already have K constraints in K unknowns. So, no need to use Lagrange multipliers at all. Instead, pick a constant $\rho > 0$ (the relative importance assigned to the constraints). Multiply the constraints by ρ, and add to the elementary (unconstrained) energy-minimization system

$$Dw = 0,$$

yielding

$$Dw + \rho w = \rho F b.$$

Finally, solve this for w. You then get your solution: $F^* w$. This is indeed smooth (has low energy) and approximates b rather well, as required.

How to optimize ρ? By trial and error. $\rho \to 0$ is too small: the constraints are nearly ignored. $\rho \to \infty$, on the other hand, is too large: it hardly minimizes energy at all. Use it only if you don't know D at all.

For a start, pick some moderate ρ. Solve for w, and look at the error: $\|F^* w - b\|$. Is it small enough? If not, then update ρ (increase or decrease it a little), and restart, until you are happy. Better yet, optimize ρ by the golden-ratio algorithm (Chapter 1 in Ref. [24]). On the way, drop noisy constraints (as at the end of Section 1.4.1). In the end, test on a lot of new data: new x_n's and a new b. If you are still unhappy, restart all over again.

5.3 Toward Image Processing

5.3.1 *Application in Image Processing*

In image processing, we work in 2-D: $d = 2$ and $x_n \in \mathbb{R}^2$ (the pixels). This way, N is often a multiple of 2^{20} (for $2^{10} \times 2^{10}$ images). The b_n's are the gray-levels: integer numbers between 0 and 255, telling us how bright the pixel is. In a color image, on the other hand, there are three basic colors, so N is a multiple of $3 \cdot 2^{20}$.

So, our Fourier transform is now doubled: a horizontal one, followed by a vertical one on top. This way, our smooth Fourier modes (of low energy) should be smooth in both directions: horizontal and vertical alike.

5.3.2 *Strong Diffusion along Edges*

So far, our A discretized the Laplacian:

$$A \doteq -\triangle + I.$$

But this is not a must. A could be any other Hermitian matrix. Likewise, F could be a new unitary matrix (not necessarily the Fourier matrix). For example, A could depend on the image itself: it could discretize a new differential operator, with variable diffusion: strong along each inner edge (interface between different objects in the image) and weak in the perpendicular direction (across the edge: Chapter 10). For this new A, pick K nearly singular eigenvectors (of a small eigenvalue) and plug them as columns in our new F^*. How should they look like? They should be nearly constant along the edge and oscillate across it (to allow a jump). This should give them a rather low energy (with respect to our new A). After all, across the edge, the diffusion is only weak and can't "see" the jump.

5.4 Toward Wavelets

5.4.1 *Local Cosines*

How to do this automatically, without even knowing A explicitly? Use local cosines (as in spectral elements): shifted and centered at any pixel. To extend them to 2-D, use a tensor product. This will allow some local oscillation (either horizontal or vertical), from either side of each pixel.

Better yet, use wavelets: pick a kernel (say, the real Gaussian) and center it around any pixel (expectation), with any scale or width (variance), and even with alternating signs (to allow high oscillations too). To extend them to 2-D, use a tensor product. These wavelets should then be nearly orthogonal to each other and can even follow edges: either horizontal or vertical or even oblique [2,3]. In the end, pick only those wavelets that are prominent in the image (have a big coefficient). As a matter of fact, local cosines can also be viewed as a special kind of wavelet.

5.4.2 *Toward Pattern Recognition and Compression*

After dropping noisy constraints and small $|w_n|$'s, we are left with only a few w_n's to help characterize the original image. As a matter of fact, these make a compressed image: up to K parameters, easily sent over the Internet. This way, the receiver can easily convert them back to a rather good image [20].

Moreover, in pattern recognition too, we can now compare two images in terms of their parameters. If they (nearly) match, then they must be just two photos of the same object or person. This is indeed AI for you.

5.5 Exercises

5.5.1 *Domain Decomposition*

1. How to design the columns in F^*? *Hint*: use the original image to introduce a new diffusion P (Chapter 10 in Ref. [24]): strong along each inner edge in the image and weak in the perpendicular direction (to allow a jump). Use P to define the denoising matrix (see there). Pick K nearly singular eigenvectors (of a small eigenvalue), and plug them as the new columns in the new F^*.
2. How to approximate these eigenvectors well? *Hint*: use a Lanczos-type iteration [28].
3. How expensive is this? *Hint*: too expensive.
4. But we only need them locally, not globally. How to do this cheaply? *Hint*: see the following.
5. For this, use domain decomposition: split your digital image into patches (subsquares), with a little overlap (Figure 5.2). For example, in an image of $2^{10} \times 2^{10}$ pixels, use $2^6 \times 2^6$ patches, of 20×20 pixels each.
6. Drop those patches of little (relative) variance, which probably contain no interesting features.

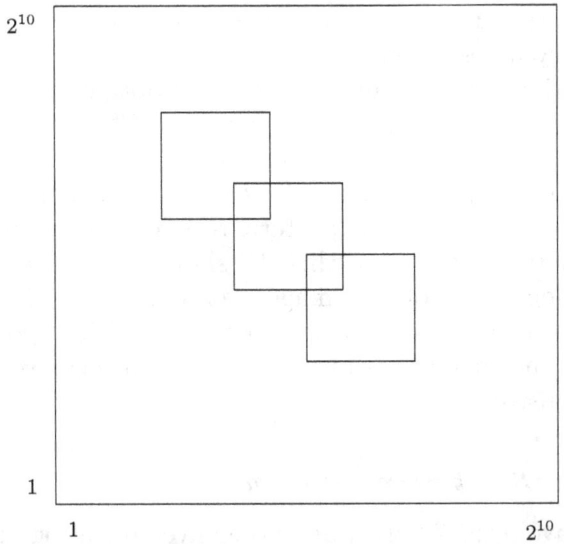

Figure 5.2. Domain decomposition: a few (overlapping) patches. In each, pick a few (local) eigenvectors: smooth along the inner edge and oscillating across it only (to allow a jump).

7. In each (remaining) patch, define the diffusion coefficient P, and use it to design the local submatrix (a block from the global matrix). *Hint*: in our example, this is a 400×400 submatrix.
8. Modify its main diagonal: for those pixels that remain outside the patch, throw their off-diagonal elements onto the main diagonal.
9. Why is this good? *Hint*: on the interface between patches, this will model homogeneous (inner) Neumann conditions (as in domain decomposition and multigrid).
10. In each patch, there are still two possibilities: if P is nearly constant, then the submatrix is much like $-\triangle$, so local Fourier modes are good enough there.
11. Extend them with dummy zeroes, and obtain a few column vectors, ready to be plugged into your new F^*.
12. If, on the other hand, P varies a lot in the patch, then there must be an interesting feature there (and an inner edge). To capture it, apply a Lanczos-type iteration to this submatrix, and obtain a few (local) nearly singular eigenvectors (of a small eigenvalue).

13. Look at them: are they indeed smooth along inner edges and oscillatory across them only?
14. Extend them with dummy zeroes, and obtain more column vectors, ready to be plugged into your new F^*.
15. Compare this to wavelets [2,3].
16. What are the pros and cons? *Hint*: our algorithm depends on the image too much. In pattern recognition, we need to compare any two images. For this, F^* should better be more general (and independent of the image), as in wavelets. Moreover, how to send a compressed image? In our algorithm, you must send not only the coefficients but also the (local) eigenvectors. This is too expensive.

5.5.2 *Use Object Segmentation*

1. How to avoid this? Forget about eigenvectors altogether. Instead, use object segmentation (Chapter 10 in Ref. [24]) to obtain a (closed) curve around the object in the original image. Design a few local cosines that are smooth along the curve, and oscillate inward only (perpendicular to the curve and into the object itself). Extend them with dummy zeroes. These will be the columns in your new F^*.
2. Is this better? *Hint*: this still depends on the image but not too much.
3. How to use this in image compression? *Hint*: no need to send F^* any more; just store the curve (bit by bit), and send it (along with the coefficients). This way, the receiver will be able to do the same: reconstruct F^* (in the same way), and use the coefficients to recover a rather good image.
4. Better yet, along with the coefficients, send the corresponding (local) cosines, compressed into a few numbers only: center, scale, and two wave numbers (horizontal and vertical).
5. Compare this to the simplest wavelet: the Haar transform [20]. Which one is better?
6. How to use the above algorithm in pattern recognition? *Hint*: compare two images, coefficient by coefficient. If they (nearly) match, then use our original algorithm to make sure that the original images indeed match.

7. How to implement the discrete Fourier transform efficiently? *Hint*: use FFT (Chapter 13).
8. Run your FFT code. How efficient is it?
9. Implement FFT nonrecursively. *Hint*: see the following.
10. Run your code. Is it any faster?
11. Parallelize this on a hypercube. The solution follows from Chapter 14.
12. Is the speedup good?

Part II: Entropy and Entanglement

What is entropy? It tells us how much uncertainty there is in our (closed) physical system. The higher the entropy, the higher the uncertainty, and the more has the system to learn yet. This also inherits to dynamics: as time goes by, information can only leak out. This is like a student who can only forget and have more and more entropy: more room to learn more. This is the second law of thermodynamics: entropy can only increase, not decrease!

This also inherits to computing and programming. In this case, we are the system (or the student). Look at a computer code: say, an if-else tree. The more even it is, the higher its entropy. Why? Because it can teach us much and give us a lot of new information. Indeed, at each node, it tells us which branch to pick and which to reject. A more uneven tree, on the other hand, can't teach us so much. It has little effect and little entropy. After all, we already knew a lot, even without it.

In the extreme case, a unary tree tells us nothing new. This is why it has zero entropy: we already know all and don't need it at all. It can't teach us anything and don't affect us in any way. This is why entropy is so useful in information theory.

Thus, the higher its entropy, the more can the code teach us and benefit us. In quantum mechanics, this takes a new face: entanglement entropy. This tells us how much information can flow from one random variable to another. The higher the entanglement entropy, the more information can flow from one measurable to another and have a real physical effect on the latter (and on its distribution). If, on the other hand, there is no entanglement entropy, then we have a complete disentanglement: the measurables don't care about each other and can never affect one another.

Disentanglement is quite boring: each random variable "lives" alone and doesn't "know" about the other. Entanglement, on the other hand, is much more interesting. For example, look at two entangled particles (with a high entanglement entropy). These may open new horizons toward a new fascinating field: quantum computing. This is still theoretical: to this day, nobody built a quantum computer yet. But we can still imagine one and prepare for it: design advanced quantum algorithms, ready to run on it. Later on, we'll see a few exciting examples.

Chapter 6

Entropy in Physics and Computing

Consider a (closed) physical system: say, a closed box, full of (warm) water. What is entropy? Well, entropy tells us how uniform the system is: to what extent all molecules have the same (kinetic) energy.

As time goes by, energy approaches its minimum: a stable (steady) equilibrium. In the process, the system gets more and more "boring": the molecules get more and more similar (with nearly the same energy). Why? Because heat tends to spread out, equally in the entire box (the heat equation: Section 1.2.1). This is why your coffee cup gets colder (like its surrounding), not hotter!

This is the second law of thermodynamics: in a closed system, entropy can only increase but never decrease! Nature favors maximum entropy, balance, and indeed boredom. No particular spot should be special or different.

Why? Because this would require more information: to specify this spot and distinguish it from the rest of the box. But more information also means more energy and effort. Why waste valuable resources on this? Our system would rather lose (or leak) information, not gain! This is how the second law of thermodynamics governs nature. Later on, we'll see this principle in computer science too.

6.1 Random Variable and Its Probability

6.1.1 *Throw a Coin*

Let's play a game: throw a coin. At probability α, it would show a head. At probability β, on the other hand, it would show a tail.

Clearly, these probabilities must sum to 1:

$$\alpha + \beta = 1.$$

After all, the coin must show something: either a tail or a head. This must happen, at probability 1. In fact, the coin can't decline: it must show tail or head. This is deterministic.

Is this a fair coin? If it is, then it is equally likely to show head or tail:

$$\alpha = \beta = \frac{1}{2}.$$

This gives us a discrete probability function (or distribution), telling us how likely we are to get head or tail (Figure 6.1).

6.1.2 *Throw a Dice*

Now, let's play another game: throw a dice. What would it show? Well, we can predict: at probability 1/6, it would show 1. At probability 1/6, it would show 2. At probability 1/6, it would show 3, and so on. Again, the probability is discrete and uniform (Figure 6.2).

6.1.3 *Random Variable*

This is a random variable: we don't know what value it would take. It could take either 1 or 2 or 3 or 4 or 5 or 6.

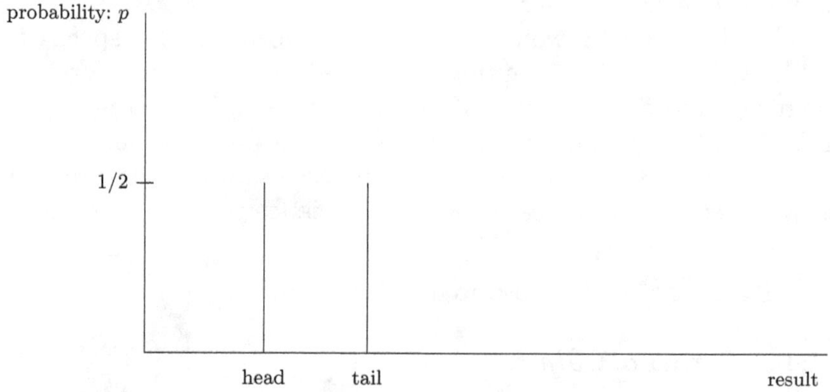

Figure 6.1. Throwing a (fair) coin: at probability 1/2, it would show a head. At probability 1/2, it would show a tail. Don't get confused: in this chapter, p will stand for probability, not momentum. (Sorry about that: this is the convention.)

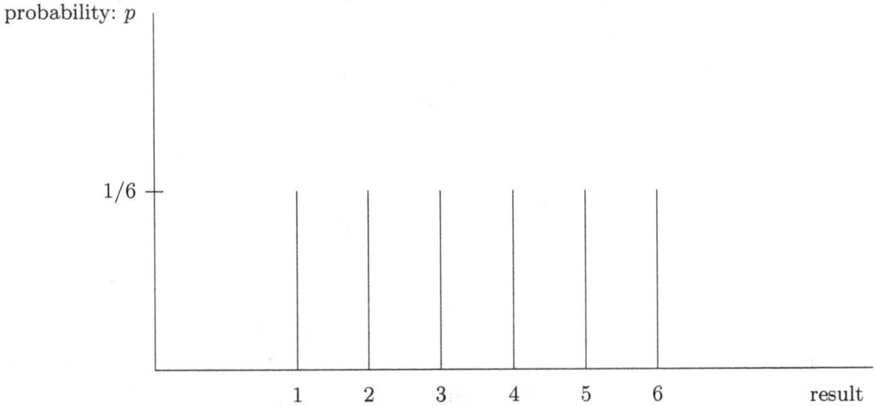

Figure 6.2. Throwing a dice: at probability 1/6, it would show 1. At probability 1/6, it would show 2. At probability 1/6, it would show 3, and so on.

If this is a good coin, then these values are equally likely. The probability is uniform: either value could be obtained at the same probability: 1/6.

6.2 Continuous Probability Function

6.2.1 *Probability and Its Distribution*

The probability function in Figure 6.2 is discrete: it is defined on six points only. This means that only six results are possible and have a positive probability. All the rest, on the other hand, are impossible: have zero probability.

More complicated random variables, on the other hand, could take infinitely many results and even a continuum of potential results: say, every result from 0 to 1. The probability function (or probability density, or distribution) should then be a (nonnegative) integrable function, defined on the entire interval $[0, 1]$. Its integral there must be 1:

$$p([0,1]) = \int_0^1 p(x)dx = 1.$$

(Don't get confused: here, p stands for probability, not momentum. Sorry about this: this is just a convention.) Why should this integral be 1? Well, integral 1 means certainty: at probability 1, the random variable must take some result between 0 and 1. This is deterministic.

How to use the probability function p? Use it as density to help integrate. Given two potential results $0 \le a < b \le 1$, the probability to obtain a result between a and b is

$$p([a,b]) = \int_a^b p(x)dx.$$

6.2.2 *Null Set: Zero Probability*

In particular, how likely are we to get the result a? This could never happen! In other words, this is quite unrealistic:

$$p([a,a]) = \int_a^a p(x)dx = 0.$$

After all, the degenerate interval $[a,a]$ is a set of zero measure. We're now ready to move on to a new concept: entropy.

6.3 Distribution and Its Information

6.3.1 *A New Random Variable*

Now, let's work the other way around: assume that we are given a probability function — a positive integrable distribution $p(x)$. Could we define an interesting random variable from scratch? Let's try

$$\log_2\left(\frac{1}{p}\right) = -\log_2 p.$$

This is a new (nonphysical) random variable: it exists in our mind only. What is its meaning? Well, it tells us how rare a certain result is. For example, look at some potential result x (of the original random variable). It has probability $p(x) > 0$. Assume that x is rare, so $p(x)$ is small, and $1/p(x)$ is large.

6.3.2 *Surprise!*

So, if we get x, how surprised should we be? As surprised as if we threw a coin many times, and always got a head, and never a tail!

How many? As many as $-\log_2 p(x)$. Indeed, the probability to get a head so many times is as small as

$$\left(\frac{1}{2}\right)^{-\log_2 p(x)} = 2^{\log_2 p(x)} = p(x)$$

(which is the probability to get x).

6.3.3 *Information*

What is the meaning of this new random variable? Well, it is quite fictitious and nonphysical. After all, it comes from our original distribution $p(x)$, which is only mathematical, not physical. Still, it has an important job: if we get x (in our original random variable), then it tells us how surprised we should be.

In particular, if we get a rare x, then we should be really surprised. How surprised? As surprised as $-\log_2 p(x)$. This is also the amount of information encapsulated in this news: we got x! The rarer x is, the smaller $p(x)$ is, and the more information we get from this news.

6.4 Von Neumann's Entropy

6.4.1 *Information and Its Expectation*

Still, our new random variable could be quite rare. Thus, to be fair, weight it by its own likelihood: $p(x)$. This will be more fair and balanced. Indeed, common x's will get a high weight, as they deserve. Even though they contain little surprise (or information), they are still common. To get prominent, they should have a big weight: $p(x)$.

This is fair now and gives us what we really want. Indeed, thanks to integration, we can now have the weighted average (expectation) of the information in our original random variable. How? Take each potential surprise (amount of information), and multiply it by its suitable weight (its probability p). Upon integrating over all possible x's, this is the expected information:

$$-\int p(x) \log_2 p(x) dx.$$

This is entropy.

6.4.2 *Von Neumann's Entropy*

This is von Neumann's entropy. It tells us a global thing: not only the amount of information encapsulated in a particular x but also the information we can expect in advance (even before carrying out an experiment to help measure and pick a specific x). After all, before the experiment, we don't know what result we'd get. In theory, we could get just any x, at probability $p(x)$. Thus, to have a fair estimate of the expected information, we must multiply x's surprise by x's probability and integrate over all possible x's, as in the above integral.

6.4.3 *Zero Probability*

Here, one may ask: what if x is impossible, so

$$p(x) = 0?$$

After all, this is an impossible case, with an infinite surprise:

$$-\log_2 p(x) = -\log_2 0 = \infty.$$

How to avoid this? Fortunately, in the entropy, we're only interested in the product

$$-p(x)\log_2 p(x) = -\frac{\log_2 p(x)}{1/p(x)}.$$

As $p(x) \to 0$, this is denoted by ∞/∞. To calculate the limit, use L'Hopital's rule: differentiate both numerator and denominator with respect to $p(x)$:

$$\lim_{p(x)\to 0} \frac{\log_2 p(x)}{1/p(x)} = \lim_{p(x)\to 0} \frac{1/p(x)}{-1/p(x)^2} = -\lim_{p(x)\to 0} p(x) = 0.$$

This could be denoted by

$$0 \cdot \log_2 0 = 0.$$

Thus, such x's contribute nothing to the above integral.

6.5 Application in Computer Science

6.5.1 *Socrates: I Know Nothing!*

Socrates said, I only know I know nothing! Entropy is quite good for our education: it tells us how little we know and how ignorant we really are. The higher the entropy, the less we know, and the more open we are to new information. What is maximum entropy? This is at a completely uniform distribution: we know absolutely nothing about our random variable and welcome any clue about what it is more likely to be!

Later on, in quantum mechanics, we'll see entropy in yet another face: entanglement entropy. Outside our system, assume that there is yet another random variable that can affect our original random variable from the outside. How much effect can it have? In other words, how much information can flow in and affect us? This is measured by the entanglement entropy. Let's see yet another interesting application: in data science, information theory, and indeed AI.

6.5.2 *Boolean Expression: Binary Tree*

In computer science, entropy takes yet another face: it characterizes a computer code and estimates how many new data it could give us. For example, in a computer program, look at an if-else tree. The more balanced it is, the more it can tell us. Indeed, at each node, it advises us which branch to pick. This is maximum entropy: a lot of new information (and indeed surprise). A less balanced tree, on the other hand, has less entropy: on average, it teaches us less. What is the most unbalanced tree? A trivial (unary) tree has zero entropy. Why? Because it tells us nothing. After all, at each node, there is just one branch to pick.

How to see this geometrically? In your tree, draw a new line, to connect the leaves, one by one. This will illustrate a distribution p back again (in a logarithmic scale). Let's see a few examples. For this, let's look at things in a wider context.

What is an if-else tree? Logically, this is actually a Boolean expression, say

$$(a \vee b) \vee (c \vee d)$$

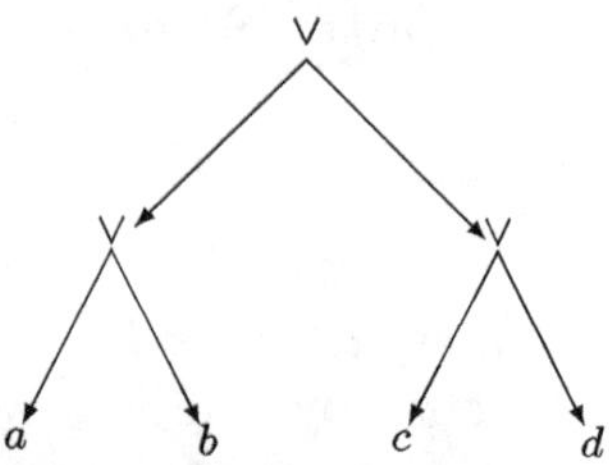

Figure 6.3. A balanced Boolean expression: $(a \vee b) \vee (c \vee d)$. Here, a, b, c, and d have the same status: either could be reached at the same probability: $p = 1/4$, giving us surprise (or new information) of $\log_2 1/p = \log_2 4 = 2$. (This is why they are at level 2.) This gives maximal entropy: $2/4 + 2/4 + 2/4 + 2/4 = 2$.

(Figure 6.3). This is highly balanced: it makes a perfect binary tree, with full levels. At the bottom, in particular, all leaves have the same status and occupy the same level: the second level. These are our random variables: the Boolean variables a, b, c, and d. Only when the code runs will they take a concrete value: either 0 (false) or 1 (true). This way, the entire expression is a random variable as well: it could be either 0 or 1 too.

What is its entropy? To get a geometrical sense, draw a new line, to connect a to b to c to d. Underneath, you'll see a new (uniform) distribution p (like those in Figures 6.1 and 6.2). In it, all Boolean variables could be reached at the same probability: 1/4. Thus, each gives us the same amount of new information (and surprise):

$$-\log_2 p = -\log_2 \left(\frac{1}{4}\right) = \log_2 4 = 2.$$

This is why, in Figure 6.3, a, b, c, and d are in level 2. We are now ready to calculate the entropy (as a weighted average):

$$\text{entropy} = \sum_{a,b,c,d} \text{surprise} \cdot \text{probability} = \sum_{a,b,c,d} 2\frac{1}{4} = \frac{2}{4} + \frac{2}{4} + \frac{2}{4} + \frac{2}{4} = 2.$$

Is this good? Well, let's compare this to a less balanced tree.

6.5.3 *An Unbalanced Boolean Expression*

Compare this to a less balanced expression:

$$a \vee (b \vee c) \vee d.$$

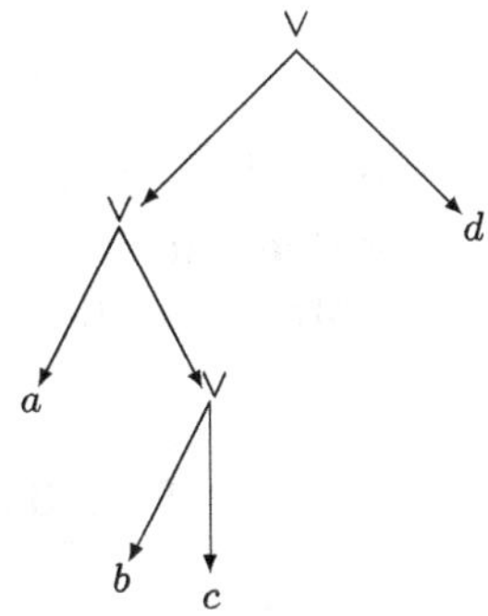

Figure 6.4. An unbalanced Boolean expression: $a \vee (b \vee c) \vee d$. At the bottom, b and c could be reached at probability 1/8 each. Therefore, they give us more surprise and information: $\log_2 8 = 3$. (This is why they are at level 3, not 2.) a, on the other hand, is as before: probability 1/4 and surprise 2. (This is why a is still at level 2.) Finally, d has probability 1/2 and surprise 1. (This is why d is at level 1.) Thus, the entropy is $6/8 + 2/4 + 1/2 = 1.75$.

In what sense is this unbalanced? To see this, draw its (unbalanced) binary tree (Figure 6.4). At the bottom, its leaves are at different levels. To picture this better, draw a line, connecting a to b to c to d. This line is no longer horizontal but broken. Underneath, you'll see a new (nonuniform) distribution p: more interesting and informative than before.

In particular, b and c are deep in level 3 (surprise: 3). a, on the other hand, is still at level 2 (surprise: 2). Finally, d is even higher: at level 1 (surprise: 1). Now, the entropy is the expected surprise (information from the code):

$$\text{entropy} = \frac{3}{8} + \frac{3}{8} + \frac{2}{4} + \frac{1}{2} = 1.75.$$

This is less than before. Why?

6.5.4 *Computer Code and Its Entropy*

What is a Boolean expression? Think of it as a computer code: each "∨" stands for an if-else question. On average, the former code is expected to tell (and surprise) us more than the latter. This is why it has more entropy.

6.5.5 *Tabula Rasa?*

How to see this geometrically? Well, the former tree was perfect: underneath the leaves in Figure 6.3 we had a uniform distribution: a Tabula Rasa that welcomes maximum information. This is top entropy. In Figure 6.4, on the other hand, underneath the leaves, the distribution is not uniform any more. This means that we are not so ignorant as before: now, we know something in advance and can't take so much new information as before. This is why the code has less entropy.

Later on, we'll see this in quantum mechanics too: the higher the entropy, the more information can flow right into the system, from the outside. How to let the new information in? Run your code! (Or, in physics, carry out an experiment.)

6.6 System and Its Entropy

6.6.1 *System of n Coins*

How to design a big system? For this, we need an elementary ingredient: a bit, or indeed a coin. Consider a system of n (independent, fair) coins: each equally likely to show 0 (head) or 1 (tail). As a whole, the entire system shows a complete configuration of 0's and 1's, say

$$011001110101010000011110000001111010101011101001.$$

6.6.2 *Configuration of Bits*

This is a list of n bits: 0 or 1. How likely are we to see this particular configuration? Well, this is quite rare: the probability for this is as low as 2^{-n}.

Is this a legitimate distribution? To check on this, let's sum up. There are as many as 2^n different configurations. Thus,

$$\sum_{\text{all configurations}} 2^{-n} = 2^n 2^{-n} = 1,$$

as required.

6.6.3 *System and Its Entropy*

What is the von Neumann entropy of the system? Well, our system is discrete, not continuous. Therefore, we need to sum, not integrate. For this purpose, scan all possible configurations, one by one. Each will contribute a product: probability times surprise. Fortunately, this is uniform — all contributions are the same:

$$\begin{aligned}
\text{entropy} &= -\sum_{\text{all configurations}} 2^{-n}\log_2\left(2^{-n}\right) \\
&= -2^n 2^{-n}\log_2\left(2^{-n}\right) \\
&= -\log_2\left(2^{-n}\right) \\
&= \log_2\left(2^n\right) \\
&= n.
\end{aligned}$$

Thus, the bigger the system, the bigger its entropy. Why? Because it has more room for new information (from the outside). How to use this in physics?

6.7 Application in Thermodynamics

6.7.1 *Closed System: Box of Gas*

How to use this in thermodynamics? Consider a box of gas (of a fixed volume: V). Assume that this is closed: no force or energy comes in or leaks out. Is this similar to a system of coins? It sure is. Instead of coins, we now have molecules of gas. Like the coins, the molecules are independent of one another. Why? Because each moves randomly, at no favorite speed or direction.

In theory, we could actually set $n = 1$ and obtain an elementary "system" of just one molecule. Still, thanks to our assumption, this would make a closed system in its own right: completely independent of its neighbors. Although they keep bumping into it, this is quite random and cancels out.

Thus, in principle, molecules are the same as coins. They are just a little more complicated: they are not static but dynamic: they move (in 3-D) and have kinetic energy. At each individual time, each molecule can move at velocity

$$v = \|v\| \cdot \frac{v}{\|v\|} \in \mathbb{R}^3.$$

This product is geometrical-physical: the speed $\|v\|$ times the direction: $v/\|v\|$.

6.7.2 *Random Variable: Kinetic Energy*

What is the kinetic energy of one individual molecule? This is

$$\text{kinetic energy} = \text{mass} \cdot \frac{\|v\|^2}{2}.$$

This depends on speed only, not direction. Besides, all molecules have the same mass. Thus, the probability to have a certain kinetic energy can be written as $p(\|v\|)$ (for all molecules alike). To serve as a legitimate probability function, it must have integral 1:

$$\int_0^\infty p(\|v\|)d\|v\| = 1.$$

(More precisely, this is some other constant — see the following.) This is an integral over all possible speeds: $0 \leq \|v\| < \infty$. Later on, we'll also write p more directly, as a function of energy (rather than speed).

6.7.3 *Classical vs. Quantum Energy*

Quantum mechanics is nondeterministic: a particle may have a few possible energy levels at the same time (each at some probability). This is a superposition: a sum of a few legitimate possibilities. In a sense, this makes a few parallel universes; each uses one of these possible energy levels and could be entered at some probability. Before measurement, the total entropy splits among these hypothetical universes. To pick the correct universe, there is a need to carry out an experiment (Section 4.5).

Here, in classical mechanics, on the other hand, things are much more calm and prosaic. At each individual time, each molecule has its own concrete (deterministic) energy, with no ambiguity. Still, does it make sense to look at each individual molecule on its own? Well, there are too many. Unlike in quantum mechanics, we could in theory look at each and even measure its concrete energy. But this would lead us nowhere: better look at the global picture, leave the individual molecules out, and look at their energy as an (unknown) random variable, with interesting (stochastic) properties. This may give us more insight about the gas as a whole and its energy (in stochastic terms). This is not because we can't be more accurate but because we don't want to: it won't serve any purpose and won't shed any light on the system or its physics.

In quantum mechanics, on the other hand, even nature remains puzzled: she can't make her mind how much energy an individual particle really has. This is why she leaves this open: vague and uncertain. All energies are still possible, thus superposed (summed up). Only upon observation does nature decide how much energy to assign and which universe to enter at the next moment (Figure 8.2). Now, let's go back to our (classical, prosaic) box of gas.

6.7.4 *The Edge*

In nature, molecules are discrete. Still, in classical physics, they are modeled as a continuum. This way, we don't sum but integrate over the entire box (and also over all possible speeds).

Here, one may ask the following: what about those molecules near the edge of the box? After all, they are not free to move in all directions, but only inward, to keep the system closed! Shouldn't their probability depend on $v/\|v\|$ too? Fortunately, the edge is too thin: just a 2-D surface that can't affect our 3-D integral.

Besides, for each molecule leaving the box on one side, there should be another molecule entering from the other side (at the same speed but in an opposite direction, at least on average, since the process is completely random). This way, the total number of molecules remains constant, as required. Better yet, why consider just a box? Our "box" doesn't have to remain a box all the time: it could follow the gas and take its current shape, leaving no molecule outside. This way, the edge gets flexible and dynamic, and all molecules are free to move.

6.7.5 *Bit: Can It Model a Molecule?*

Can our simple bit model a real molecule? At first glance, it seems too simple and discrete; it only takes two possible values: 0 or 1. Besides, it has a constant probability:

$$p(0) = p(1) = \frac{1}{2}.$$

But don't worry: the bit is still good enough for the job (at least in classical terms). Moreover, as we'll see soon, a system of n bits (whose entropy is simply n: Section 6.6.3) is good enough to model a complete box of gas.

6.7.6 *Entropy: Integration*

Like a system of coins, our box of gas should also have entropy proportional to the total number of molecules (or indeed the volume: V). Indeed,

$$\begin{aligned}
\text{entropy} &= -\int_0^\infty \iiint_{\text{box}} p(\|v\|)\log_2\left(p(\|v\|)\right) d\|v\|dxdydz \\
&= -\int_0^\infty p(\|v\|)\log_2\left(p(\|v\|)\right) d\|v\| \iiint_{\text{box}} dxdydz \\
&= -V\int_0^\infty p(\|v\|)\log_2\left(p(\|v\|)\right) d\|v\|.
\end{aligned}$$

Is this familiar? This is just a product: the volume times the entropy of one molecule. In particular, this is proportional to the volume, as required. This shows once again how suitable our coins are to model thermodynamics.

6.8 Systems: Are They Additive?

6.8.1 *Two Systems and Their Tensor Product*

What did we do with our coins? We started from one coin (or bit): as simple as you can get. To this, we added another bit. To this, we added yet another bit, and so on, to make a complete system of n bits. Later on, this will be useful in quantum computing too.

Algebraically, this is a tensor product (repeated, time and again). How to generalize this? Start from two independent systems. On one, define some random variable, with value x. (In a bit, for example, x can only be 0 or 1. In a real random variable, on the other hand, $x \in \mathbb{R}$.) The probability to obtain x is, say, $p_1(x)$.

On the other system, define an independent random variable: y. The probability to obtain y is, say, $p_2(y)$. How to design a joint system, with a joint random variable? Take their tensor product: a pair (x, y). The probability to obtain it is

$$p(x, y) = p_1(x)p_2(y).$$

This is a tensor product again.

6.8.2 *Separability: Independence*

This decomposition means that p is separable. Why? This is thanks to the original relation between x and y: independence. Later on, in quantum mechanics, we'll see an even more powerful kind of separability: in not only a distribution but also a wave function (Section 8.3.2). This (strong) separability will lead to our (weak) separability and indeed to independence (but not the other way around).

6.8.3 *Independent or Disentangled?*

How is classical mechanics different from quantum mechanics? In classical mechanics (and indeed thermodynamics), we have no wave function but only probability. Our joint probability p has a new property: separability. Why? Because the original random variables are independent of one another. This also works the other way around: separability leads to independence back again. So, separability and independence are actually one and the same.

In quantum mechanics, on the other hand, we'll see an even stronger property: disentanglement. This means even more: even the joint wave function is separable. Square it up, and get your joint (separable) distribution too. Thus, disentanglement is quite powerful: it means separability of both wave function and distribution (leading to independence too).

What about the other way around? Does independence imply disentanglement? No! Disentanglement is just too powerful! For example, in a coherent state, position and momentum are only independent but not disentangled (Section 2.5.7).

Here, however, we are still in classical physics. Here, things are still calm and easy: there is no wave function at all (hence no interference either). So, separability must refer to the joint distribution and be equivalent to independence.

So, let's look at our joint probability again. Is this a legitimate probability function? Let's check: it is indeed nonnegative and has integral 1:

$$\iint p(x,y)dxdy = \iint p_1(x)p_2(y)dxdy = \int p_1(x)dx \int p_2(y)dy = 1{\cdot}1 = 1.$$

Next, in our joint system, what is the entropy?

6.8.4 *Entropy Is Additive*

Let's calculate the entropy of the joint system:

$$\begin{aligned}
\text{entropy} &= -\iint p(x,y)\log_2\left(p(x,y)\right)dxdy \\
&= -\iint p_1(x)p_2(y)\log_2\left(p_1(x)p_2(y)\right)dxdy \\
&= -\iint p_1(x)p_2(y)\left(\log_2\left(p_1(x)\right)+\log_2\left(p_2(y)\right)\right)dxdy \\
&= -\iint p_1(x)p_2(y)\log_2\left(p_1(x)\right)dxdy \\
&\quad -\iint p_1(x)p_2(y)\log_2\left(p_2(y)\right)dxdy \\
&= -\int p_2(y)dy\int p_1(x)\log_2\left(p_1(x)\right)dx \\
&\quad -\int p_1(x)dx\int p_2(y)\log_2\left(p_2(y)\right)dy \\
&= -\int p_1(x)\log_2\left(p_1(x)\right)dx - \int p_2(y)\log_2\left(p_2(y)\right)dy.
\end{aligned}$$

What is this? This is just the sum: the entropy of the first system, plus the entropy of the second one. Thus, entropy is additive. How to use this in practice?

6.8.5 *Back to System of Coins*

How to use additivity? Well, we already used it (implicitly) in a system of coins. Indeed, our original coin (or bit) is quite fair (has a uniform probability): it has just two possible values, at probability 1/2 each. This is why it has entropy 1:

$$-\frac{1}{2}\log_2\left(\frac{1}{2}\right)-\frac{1}{2}\log_2\left(\frac{1}{2}\right)=\frac{1}{2}\log_2 2+\frac{1}{2}\log_2 2=\frac{1}{2}+\frac{1}{2}=1.$$

Now, add another coin. This makes a system of two independent coins. Thanks to additivity, it has entropy 2. Next, add yet another coin, to make a system of three independent coins, with entropy 3, and so on. By mathematical induction, a system of n coins will indeed have entropy n, in agreement with Section 6.6.3.

6.8.6 *Back to Box of Gas*

In computer science, the bit is the unit of information. Why? Because it is elementary and has entropy 1. Thanks to additivity, this also extends to the whole computer: its entropy is as high as the total number of bits in it.

In thermodynamics, on the other hand, the elementary unit is the molecule. Still, in principle, this is just the same as before: thanks to additivity again, in our box of gas, the total entropy is a product — the volume (total number of molecules) times the entropy of one individual molecule. Still, why volume?

In real nature, gas is discrete: made of many discrete molecules. Still, we often model it as a continuum (in 3-D). In this model, our probability is no longer discrete but continuous: a new probability density (or distribution). To calculate the entropy, we must now integrate, not sum. This is why, in Section 6.7.6, we indeed integrated over the entire box. This gave us the entropy of the entire box as a product: volume times the entropy of one molecule. This is additivity again, in a continuum.

6.9 Laws of Thermodynamics

6.9.1 *First Law: Energy Is Conserved*

What is a closed system? Well, on a closed system, no force acts from the outside. Thus, a closed system is isolated (in terms of forces).

In a closed system, energy is conserved: remains the same, all the time. This is the first law of thermodynamics: the total energy is constant, with no change over time.

6.9.2 *Second Law: Entropy Increases*

Thus, in a closed system, energy is fixed. Entropy, on the other hand, is not so strict. Unlike energy, entropy is not quite constant but only one-sided. In a closed system, entropy can only increase but never decrease. Why?

Well, we already know why. More entropy means that our random variable is expected to give us more surprise and news. Thus, *before* running an experiment to measure the random variable, there was a rather uniform distribution, with a lot of room for new information.

Thus, more entropy means more uncertainty: more uniformity in the original distribution. This is why, in nature, entropy can only increase, not decrease (unless an experiment is carried out). In other words, the system can't gain information from nothing but only lose: get more and more disordered and vague. Nature seeks maximum entropy (and disorder and indeed uncertainty): a distribution as unfocused (and spread-out) as possible. This follows from in the heat equation (Sections 1.2.1 and 1.2.2). This is the second law of thermodynamics.

6.10 Maxwell's Demon

6.10.1 *Example: Cup of Coffee*

How to see this in a concrete example? Consider a cup of coffee. Assume that it is completely isolated: no heat leaks from it to the air. This makes a closed system.

6.10.2 *Entropy Increases*

Initially, the coffee is hot, and the cup itself is cold. This means low entropy: the system must contain a lot of data to distinguish between the coffee and the cup and tell us that only the coffee is hot but the cup is not.

As time goes by, the coffee gets colder: heat flows from the coffee to the cup, which gets warmer. As a whole, our system gets more and more calm, balanced, and benign and less sharp or dichotomic.

By now, the coffee is no longer homogeneous. At the bottom, near the cup itself, it is already rather cold. At the top-middle, far away from the cup itself, it is still warm. This is not as strict as in the beginning: we lost information and gained entropy. Reality is not quite black or white but mostly gray, in between. Entropy got higher, as in the second law (and indeed in the heat equation).

6.10.3 *Equilibrium: Minimum Energy — Maximum Entropy*

Nature is quite economic: she seeks minimum energy and maximum entropy. This is why the coffee gets colder, not hotter. This way, the entire system (including the cup itself) gets more and more balanced and homogeneous (in terms of heat).

Nature seeks equilibrium: minimum energy and maximum entropy (as uniform as possible). This is the (final) steady state, which can't change any more. In computer science too, an efficient tree is as balanced as possible: full and perfect, with all leaves at the same level (Figure 6.3). This could model an efficient code, giving us a lot of new information, for future use.

6.10.4 *Maxwell's Demon*

Let's do a little thought experiment. In our cup of coffee, put a thin (vertical) interface, to divide it (vertically, symmetrically) into two equal halves: the left half mirrors the right half. On the interface, at the very center, put a demon. This is Maxwell's demon, whose job is to violate the second law of thermodynamics. How? It examines those molecules that attempt to cross the interface (from left to right) and measures their speed. (Assume that this costs no energy.) Only those of high speed area allowed, while those of low speed are sent back leftward. In the end, thanks to randomality, our coffee will get half-hot-half-cold: hot on the right, and cold on the left! This has more information than in the beginning: a new strict dichotomy between hot and cold. Entropy has decreased! How come? What about the second law of thermodynamics?

6.11 Data Unit: Shannon's Bit

6.11.1 *Maxwell's Demon and Energy*

So far, we ignored an important question: where does the demon take its energy from? After all, the demon does a lot of work: examine the molecules, measure their speed, force some of them to change their momentum, and go back leftward. In information theory, this is like recording (storing) data. For this, green energy is not good enough. The demon must burn good fuel, which makes entropy jump (see exercises in the following).

6.11.2 *Data and Memory*

Maxwell's demon is an imaginary computer code (or an experiment): it tests each molecule and stores the result in the molecule itself (its place in the cup). If you like, leave it theoretical, to help estimate entropy, better yet, run it in practice, to help simulate a real computer.

To store data, you need not only energy but also memory. In information theory, the memory is an integral part of the system. Memory is valuable and quite expensive.

At some point, the demon will run out of memory and will have to remove some old data, to make room for new. This is irreversible: these old data will be lost and "forgotten" forever. Entropy will increase, as in the second law of thermodynamics.

6.11.3 *Shannon's Bit*

As a theoretical tool, Maxwell's demon helps us estimate entropy. Indeed, what is entropy in physics and computing? Entropy is the amount of missing information (that can still be completed by running a code or an experiment). The more elements in our system (and the more random they are), the more room for new data, and the higher the entropy. Thus, to estimate entropy, imagine a theoretical computer. The bigger it is (and the cleverer its software), the more information it could give us about our original system, and the higher our entropy.

Maxwell's demon, although tiny, is a big "computer," with a lot of memory: the molecules themselves. It is ready to run an experiment, with a perfect if-else question. No need to run it: it just indicates that we have high entropy from the start. How high? As high as the total number of molecules (or memory bits). After all, this is the total amount of data that our system can take.

Shannon introduced a new unit of data: a bit — 0 or 1. When your (real) computer runs a code, it may fetch data from the memory, calculate (in its processor), and store the result in the memory back again. This is still reversible: thanks to the new datum, entropy drops a little but also jumps a little (to produce the required energy), which cancels out. When does entropy increase? Only when some old datum is dropped (to make room for a new one). Only then does the system lose information or gain entropy.

Later on, in quantum mechanics, the bit will help model spin. More precisely, this will already make a qubit, which stores much more: not only 0 or 1 but also their probability: $p(0)$ and $p(1)$. How? For this, it will store even more: the wave function, $u(0)$ and $u(1)$, to help calculate $p(0) = |u(0)|^2$ and $p(1) = |u(1)|^2$. Thanks to their complex value, $u(0)$ and $u(1)$ are ready for superposition, interference, and quantum computing.

6.12 Maxwell–Boltzmann Distribution

6.12.1 *Random Variable: Energy*

How to use entropy in physics? In Section 6.7.2, we already saw a closed system: a box of gas. In such a system, the relevant random variable is (kinetic) energy. After all, molecules of the same energy are essentially the same.

In the box, the molecules move at random, all the time. Still, they have the same mass. Thus, their energy depends on speed only (not direction):

$$\text{energy} = \text{mass} \cdot \frac{\|v\|^2}{2}.$$

What is the probability to find a molecule of a certain energy? Well, let's write it as $P(e)$ (where e stands for kinetic energy). This could also be viewed as a (composite) function of speed: $p \equiv p(\|v\|)$. Still, the former style is more simple, direct, and natural. Let's use it.

6.12.2 *Maxwell–Boltzmann Statistics*

How should p look like? This is the Maxwell–Boltzmann distribution (or statistics):

$$p(e) = \frac{\exp\left(-\frac{e}{K_B T}\right)}{K_B T}.$$

This is the probability to find a molecule of energy e, under the following circumstances:

- The overall temperature is T (in Kelvin's scale).
- The average energy is, thus, $K_B T$ (where K_B is Boltzmann's constant).

Thanks to this definition, p is indeed a legitimate probability function: it is positive and has integral 1, as required.

6.12.3 *Surprise!*

What is the surprise? For simplicity, let's use the natural logarithm:

$$-\log(p(e)) = \frac{e}{K_B T} + \log K_B + \log T.$$

What do we have here? We have three terms. Look at the first one: $e/(K_B T)$. It tells us how surprised we should be to discover a molecule of energy e. The higher the energy, the lower the probability, and the more surprised we should be. After all, it is quite a (pleasant) surprise to find a fast molecule (especially at a low temperature)! Thus, the Maxwell–Boltzmann distribution indeed makes sense.

6.13 Exercises: Conservation Laws

6.13.1 *Conservation Laws in a Closed System*

1. What is a closed system? *Hint*: no force acts on it from the outside.
2. What is an open system? *Hint*: there is an outer force, acting on it from the outside.
3. For example, look at a simple universe, containing just two items: the Earth and a meteor, falling toward it.
4. Look at the meteor. What happens to it? *Hint*: it accelerates downward, toward the Earth.
5. In this process, what happens to its momentum? *Hint*: it keeps pointing downward and grows in magnitude.
6. How come? What about conservation of momentum? *Hint*: the meteor on its own is an open system. Indeed, the Earth applies an outer force to it: gravity.
7. What is force? *Hint*: force is time-derivative of momentum. This is why the meteor accelerates downward: its momentum grows downward.
8. How to fix this and get momentum conserved? *Hint*: close the system.
9. How? *Hint*: throw the entire Earth in.
10. How does the system look like now? *Hint*: two items: the Earth and the meteor. In our simple model, this is the entire universe.
11. Is this closed now? *Hint*: yes. After all, there is no "outside" any more.
12. Is momentum conserved now? *Hint*: yes.
13. Why? *Hint*: thanks to Newton's third law of nature, there is a (symmetric) reaction: the meteor attracts the entire Earth a little and gives it an opposite momentum, just enough to cancel its own new (extra) momentum.
14. Write this in a more general (differential) form. *Hint*: look at a general (closed) system: $V \subset \mathbb{R}^3$. In V, define two vector functions: momentum p and force F. (Don't confuse this p with probability.) Recall that force is time-derivative of momentum. Since the total force is zero, the total momentum is conserved:

$$\frac{d}{dt}\iiint_V p\,dxdydz = \iiint_V \frac{dp}{dt}dxdydz = \iiint_V F\,dxdydz = \mathbf{0} \in \mathbb{R}^3.$$

15. Look at our simple model again: Earth and meteor. Is energy conserved? *Hint*: thanks to the first law of thermodynamics, energy is indeed conserved.
16. What is this energy? *Hint*: total energy: kinetic energy plus potential energy, for the Earth, plus the meteor. This sum is constant (conserved).
17. How to use this in practice? *Hint*: ignore the (tiny) energy of the Earth. Now, at each individual time, solve for the height and velocity of the meteor (in terms of its initial height and velocity).
18. Imagine that you swim in the sea. Suddenly, you forget how to swim! Can you still pull yourself in the hair and save yourself from drowning? *Hint*: unfortunately not. Due to the third law of nature, your hair will apply the opposite force back on you, leading to a net force of zero back again.

6.13.2 *Green Energy: Does Entropy Jump?*

1. Consider a computer of n bits. What is its entropy? *Hint*: like a system of n coins: n.
2. On the other hand, how much information can it contain? *Hint*: n bits of information.
3. What does this mean for us? *Hint*: using the computer is like carrying out an experiment. Before doing this, we're at a deficit of n bits of information. In other words, we have entropy n.
4. Next, consider a steam engine. How does it work? *Hint*: it burns fuel, generates heat, and boils water into steam, with a lot of kinetic energy that can do work for us.
5. How can it generate heat? After all, this heat is local and focused: a lot of information, with low entropy. To get there, entropy must decrease. Isn't this in violation of the second law of thermodynamics? *Hint*: don't worry: in the beginning, entropy was even lower. Why? Because the potential (chemical) energy was quite focused: concentrated in the fuel only, not elsewhere. Once the fuel is burnt, this potential energy is gone. We get an absolute uniformity, with a considerable jump in entropy.
6. Assume that you think about a problem and find a solution. So, you gained information and lost entropy. How come? Isn't this in violation of the second law of thermodynamics. *Hint*: don't worry:

in the process, your body burned food to generate (and also spread out) heat, with a considerable jump in entropy.

7. What do you think about so-called "green" energy (solar, wind, etc.)? Does it have a chance to power the world? *Hint*: it only saves (and reuses) energy but produces none new. Indeed, there is no jump in entropy.
8. And what are its cons? *Hint*: it interferes with healthy processes in nature and may harm the environment. Unfortunately, this is disregarded in most models.
9. On the other hand, how about fuel? Can it power the world? *Hint*: at first, it consumes a little energy, to ignite. But this is a price worth paying: when it burns, its energy falls deep into a potential well, releasing a lot of potential energy, to do work for us. In the process, entropy jumps, as required.

Chapter 7

Wave Function and the Fourier Transform

The original wave function is defined on the (real) horizontal axis. This way, it is a function of position: $x \in \mathbb{R}$. Apply the Fourier transform to it, and obtain the momentum wave function, defined on the (imaginary) vertical axis, in terms of momentum: $p \in \mathbb{R}$. This is how the Fourier transform rotates the entire complex plane (counterclockwise) and maps the real axis onto the imaginary axis. From the momentum wave function, we can now deduce how momentum is distributed.

This leads to the joint position–momentum wave function, defined on the entire phase plane. Thanks to it, position–momentum mirror one another and are treated uniformly, in the same way. Indeed, restrict the joint wave function to a horizontal line, and obtain the wave function. Likewise, restrict it to a vertical line, and obtain the momentum wave function. Later on, this will help define the density matrix and the entanglement entropy.

This chapter is written in a new style: exercises, alone with their solution. This way, you can pause and try to figure the answer yourself, before reading it. This could make the theory more alive and close to heart.

7.1 Wave Function and Distribution

7.1.1 *Wave Function*

1. Let $u(x)$ be the wave function, defined for each position $x \in \mathbb{R}$:

$$u : \mathbb{R} \to \mathbb{C}.$$

2. Is u physical? *Hint*: not quite. After all, the position x can never be measured physically (unless entangled to the entire measuring device, leading to a new bigger system — see the following).
3. Still, is u a good mathematical tool? *Hint*: mathematically, u encapsulates all information about position–momentum (in 1-D). In particular, $|u|^2$ will make a distribution (probability density) for position.
4. Write u as an infinite column vector. *Hint*:

$$u \equiv \begin{pmatrix} \vdots \\ u(x) \\ \vdots \end{pmatrix}.$$

This is also a tall $\mathbb{R} \times 1$ (complex) matrix (with $x \in \mathbb{R}$ as a row index).
5. Normalize u to have norm 1. *Hint*: divide u by its own norm. Once this is done,

$$\|u\|^2 = (u, u) = \int_{-\infty}^{\infty} \bar{u}(x)u(x)dx = \int_{-\infty}^{\infty} |u|^2 dx = 1,$$

as required.
6. Thanks to this normalization, what is the probability to find the particle at x (with any momentum)? *Hint*: $|u(x)|^2$.

7.1.2 *Probability Density*

1. More precisely, this is probability density (probability per unit length):

$$|u(x)|^2 = \frac{d(\text{probability to have position} < x)}{dx}.$$

2. This is good for (1-D) integration:

$$\int_a^b |u(x)|^2 dx = \text{probability to have } a < \text{position} < b.$$

3. For this purpose, what physical unit should $|u|^2$ have? *Hint*: 1/length.
4. For this purpose, what physical unit should u have? *Hint*: $1/\sqrt{\text{length}}$.
5. How is this related to classical physics? *Hint*: see Sections 6.8.2–6.8.6.
6. Look at the position operator X. How does it act on u? *Hint*:

$$(Xu)(x) = xu(x).$$

7. Recall that position is a random variable. As such, it has no deterministic value.

7.1.3 *Expectation: Expected Value*

1. Still, X does have an expected value. What is the expected position? More precisely, what is the expectation of X at u? *Hint*:

$$(u, Xu) = \int_{-\infty}^{\infty} \| u|^2 x dx.$$

2. How to calculate this even before u was normalized? *Hint*: $(u, Xu)/\|u\|^2$. item As an (infinite) $\mathbb{R} \times \mathbb{R}$ matrix, how does X look like? *Hint*: X is diagonal:

$$X \equiv \text{diag}(x)$$

(where $x \in \mathbb{R}$ is the diagonal index and also the diagonal element).
3. In this context, how does Dirac's delta function act on u? *Hint*: it is integrated against u (as in a convolution), producing the identity operator:

$$\int_{-\infty}^{\infty} \delta(x - \tilde{x}) u(\tilde{x}) d\tilde{x} = u(x).$$

4. Write this as an inner product as well. *Hint*:

$$(\delta(x - \tilde{x}), u(\tilde{x})) = u(x)$$

(where $\tilde{x}$ is integrated upon).

5. In this integral, what is the kernel? *Hint*: $\delta(x - \tilde{x})$ (where δ is Dirac's delta function, centered at 0).

7.2 Fourier Transform and Momentum Wave Function

7.2.1 *Fourier Transform*

1. In this context, how does the Fourier matrix look like? *Hint*: F is an infinite $\mathbb{R} \times \mathbb{R}$ matrix, whose $(x, \tilde{p})$th element is

$$F_{x,\tilde{p}} \equiv \frac{1}{\sqrt{2\pi}} \exp(ix\tilde{p})$$

(where $x \in \mathbb{R}$ is the row index and $\tilde{p} \in \mathbb{R}$ the column index). This is a little different from W_n in Section 13.1.3, which picked a minus sign in the exponent. For simplicity, we assume here $\bar{h} = 1$.

2. What is its Hermitian adjoint? *Hint*: in F^*, the (p, x)th element is

$$F^*_{p,x} = \frac{1}{\sqrt{2\pi}} \exp(-ipx)$$

(where p is the row index and x is the column index).

3. How to apply the Fourier transform to u? *Hint*: apply F^* to u:

$$(F^* u)(p) = \frac{1}{\sqrt{2\pi}} \int_{-\infty}^{\infty} \exp(-ipx) u(x) dx$$

(where p is the row index and x is integrated upon and gone).

4. Write this as an inner product as well. *Hint*:

$$(F^* u)(p) = \left(\frac{1}{\sqrt{2\pi}} \exp(ipx), u(x) \right)$$

(because, in the inner product, the first function picks a complex conjugate on top).

5. Algebraically, what is the main property of F? *Hint*: F is unitary (norm-preserving):

$$F^* F = F F^* = \text{the identity operator.}$$

6. Indeed, once F^*F is interpreted as an integral operator, what is its kernel? *Hint*: apply F^* to the $\tilde{p}$th column in F. For this, integrate on x:

$$\begin{aligned}(F^*F)_{p,\tilde{p}} &= \left(\frac{1}{\sqrt{2\pi}}\exp(ipx), \frac{1}{\sqrt{2\pi}}\exp(ix\tilde{p})\right) \\ &= \lim_{r\to\infty}\frac{1}{2\pi r}\int_{-\pi r}^{\pi r} r\exp(-i(p-\tilde{p})x)dx \\ &= \delta(p-\tilde{p}),\end{aligned}$$

   where p is the row index in F^* and $\tilde{p}$ is the column index in F. (In the latter integrand, both p and $\tilde{p}$ are multiples of $1/r$, producing a periodic function.)
7. Mathematically, what is p? *Hint*: wave number.
8. Physically, what is p? *Hint*: momentum (divided by $\bar{h}$, but this is disregarded here, by assuming $\bar{h} = 1$).
9. Indeed, apply the Fourier transform, followed by its inverse. What do you get? *Hint*: change the order of integration — integrate on dp before integrating on $d\tilde{x}$:

$$\begin{aligned}(F(F^*u))(x) &= \left(\frac{1}{\sqrt{2\pi}}\exp(-ipx), \left(\frac{1}{\sqrt{2\pi}}\exp(ip\tilde{x}), u(\tilde{x})\right)\right) \\ &= \left(\left(\frac{1}{\sqrt{2\pi}}\exp(-ipx), \frac{1}{\sqrt{2\pi}}\exp(-ip\tilde{x})\right), u(\tilde{x})\right) \\ &= (\delta(x-\tilde{x}), u(\tilde{x})) \\ &= u(x)\end{aligned}$$

   (where x is the row index in F and $\tilde{x}$ is the column index in F^*).
10. What is this algebraically? *Hint*: the associative law:

$$F(F^*u) = (FF^*)u = u.$$

11. Design yet another kernel, to help represent X as an integral operator as well. *Hint*: $x\delta(x-\tilde{x})$.
12. Prove this. *Hint*:

$$\int_{-\infty}^{\infty} x\delta(x-\tilde{x})u(\tilde{x})d\tilde{x} = xu(x).$$

13. Write this as an inner product as well. *Hint*:

$$(Xu)(x) = (x\delta(x - \tilde{x}), u(\tilde{x})) = xu(x)$$

(where $\tilde{x}$ is integrated upon and gone).

7.2.2 *Momentum Wave Function*

1. Likewise, let $w(p)$ be a momentum wave function, defined for all $p \in \mathbb{R}$:

$$w : \mathbb{R} \to \mathbb{C}.$$

2. Write w as an infinite column vector. *Hint*:

$$w \equiv \begin{pmatrix} \vdots \\ w(p) \\ \vdots \end{pmatrix}.$$

This is also a tall $\mathbb{R} \times 1$ (complex) matrix (with p as a row index).
3. Normalize w to have norm 1. *Hint*: divide w by its own norm. Once this is done, we'll indeed have $\|w\| = 1$, as required.
4. Assume that you are given a wave function $u(x)$, as above. How to define a suitable momentum wave function w that agrees with it? *Hint*: for this, apply the Fourier transform:

$$w \equiv F^* u$$

(up to an outer phaseshift, which has no physical effect at all).
5. Is this still the same state? *Hint*: in this case,

$$u = (FF^*)\, u = F\,(F^* u) = Fw.$$

This means that w represents the same state, only in terms of Fourier modes (columns of F).
6. In general, thanks to w, how likely is the particle to have momentum p (at any position)? *Hint*: thanks to normalization, the probability for this is $|w(p)|^2$.
7. More precisely, this is probability density (probability per unit momentum):

$$|w(p)|^2 = \frac{d(\text{probability to have momentum} < p)}{dp}.$$

8. This is good for (1-D) integration:

$$\int_a^b |w(p)|^2 dp = \text{probability to have } a < \text{momentum} < b.$$

9. How does P act on w?
10. Design a kernel for P. *Hint*: $p\delta(p - \tilde{p})$.
11. Prove this. *Hint*:

$$\int_{-\infty}^{\infty} p\delta(p - \tilde{p})w(\tilde{p})d\tilde{p} = pw(p).$$

12. What is the expected momentum at w? *Hint*:

$$(w, Pw) = \int_{-\infty}^{\infty} |w|^2 pdp.$$

13. How to calculate this even before w was normalized? *Hint*: $(w, Pw)/\|w\|^2$.

7.2.3 *Position–Momentum Wave Function*

1. Let $v(x, p)$ be a joint position–momentum wave function:

$$v : \mathbb{R}^2 \to \mathbb{C}.$$

2. Give an example. *Hint*: say, a coherent state (Figure 2.2), "normalized" on each horizontal line (as at the end of Section 2.5.5).
3. How does v improve the visualization of the physical state? *Hint*: on each horizontal line, v may have a different momentum (oscillation). Likewise, on each vertical (issuing from x), v has a different expected position: x.
4. Does v have a physical meaning pointwise? *Hint*: no! Due to the uncertainty principle, x and p can never be measured simultaneously.
5. In this sense, is v much different from the original wave function $u(x)$, defined above? *Hint*: $u(x)$ is nonphysical too: position can't be measured, unless entangled into a bigger system that contains our measuring device as well (see the following).
6. Still, how is v useful mathematically? *Hint*: not quite pointwise but more in the sense of (2-D) integration.

7. Later on, we'll also see a more physical v. What? *Hint*: we'll replace x and p by new (commutative) random variables that can indeed be measured simultaneously.
8. Write v as an infinite column vector. *Hint*:

$$v \equiv \begin{pmatrix} \vdots \\ v(x,p) \\ \vdots \end{pmatrix}.$$

This is also a tall $\mathbb{R}^2 \times 1$ (complex) matrix, with (x,p) as a row index.
9. Normalize v to have norm 1. *Hint*: divide v by its own norm. Once this is done,

$$\|v\|^2 = (v,v) = \int_{-\infty}^{\infty}\int_{-\infty}^{\infty} |v|^2 dxdp = 1,$$

as required.
10. How likely is the particle to be at x, with momentum p? *Hint*: thanks to this normalization, the (theoretical) probability for this is $|v(x,p)|^2$.
11. Is this physical? *Hint*: no! Due to the uncertainty principle, x and p can never be measured simultaneously.
12. Still, how to use $|v(x,p)|^2$ to estimate the probability? *Hint*: not quite pointwise but only in the sense of (2-D) integration.
13. How to make v physical? *Hint*: let x belong to one particle and p to another. This way, they do commute (see the following).
14. On each vertical (issuing from any x), integrate $|v|^2$:

$$\rho(x) \equiv \int_{-\infty}^{\infty} |v(x,p)|^2 dp.$$

(Don't confuse this ρ with the relative importance used in AI.)
15. Conclude that $\rho(x) \geq 0$.
16. What is $\rho(x)$ physically? *Hint*: the probability to find the particle at x (with any momentum).
17. Is this familiar? *Hint*: in the above, this was denoted by $|u(x)|^2$.
18. Show that those x's for which $\rho(x) = 0$ are highly unlikely and can be dropped.

19. Conclude that, for every relevant x, $\rho(x) > 0$.
20. On the above vertical (issuing from x), look at the ratio

$$\frac{|v(x,p)|^2}{\rho(x)}.$$

21. What is this physically? *Hint*: for a particle at x, this is the probability to have momentum p.
22. How could this agree with the uncertainty principle that tells us that a particle found at x must have a uniform probability to have any momentum? *Hint*: this uniform probability is valid only *after* looking at the particle and finding it at x. This changes the original state forever and projects it on a completely new state. (If you like, this is already a bigger system that contains the entire measuring device for position — see the following.) But here, our v represents the original state, before looking.
23. Likewise, look at the (complex) ratio

$$\frac{v(x,p)}{\sqrt{\rho(x)}}.$$

24. What is this physically? *Hint*: for a particle at x, this is the momentum wave function (up to an outer coefficient).
25. Is this familiar? *Hint*: in the simple case considered in the beginning, this was independent of x and denoted by $w(p)$. Here, on the other hand, we have a better visualization: on each vertical (issuing from some x), we have a different expected position: x.
26. Let H be the Hamiltonian. How does it act on v (as an integral operator)? *Hint*: assume that H has a kernel (denoted by H too):

$$H : \mathbb{R}^4 \to \mathbb{C}.$$

Thanks to this,

$$(Hv)(x,p) = \int_{-\infty}^{\infty} \int_{-\infty}^{\infty} H(x,p,\tilde{x},\tilde{p}) v(\tilde{x},\tilde{p}) d\tilde{x} d\tilde{p},$$

where (x,p) is the row index and $\tilde{x}$ and $\tilde{p}$ are integrated upon and gone.
27. How to make H diagonal? *Hint*: in H, let X^2 act horizontally and P^2 vertically.

Chapter 8

Entanglement Entropy

Thanks to the joint position–momentum wave function, we are now ready to define the density matrix (for momentum alone). From it, we'll also define the entanglement entropy. This will tell us how much information flows from position to momentum, and even affects momentum, and its distribution. We'll also see an interesting example, where position affects energy: not directly but only indirectly (through its implicit effect on momentum and its distribution).

In one particle, however, position–momentum don't commute. Thus, they can never be measured simultaneously. This is the uncertainty principle. Thus, our joint position–momentum wave function is a bit too theoretical. How to make it more physical? Look at two particles. Position belongs now to one and momentum to the other. (Or, if you like, pick any two measurables.) This way, they do commute and can be measured simultaneously.

If they are entangled (correlated), then they affect one another. This is marked by their positive entanglement entropy. If, on the other hand, they are disentangled, then they don't care about one another at all. They have no entanglement entropy and no effect on (the distribution of) one another.

These measurables don't have to be position–momentum. For instance, they can be spin. This leads to Bell's experiment that helped establish that nature is inherently stochastic and nondeterministic.

Finally, we use set theory to extend the double-slit experiment. This experiment uses a screen with two (parallel) slits in it. Still, behind the screen, you can go ahead and place infinitely many screens, one by one in a row, each with two slits in it. Thanks to the axiom of choice, you can then show how an electron (or photon) can still go through all of them, one by one, and interfere at infinity. This is probably how nature really works.

8.1 Density Matrix

8.1.1 *Density Matrix*

1. So far, we looked at v. Next, what is v^*? *Hint*: since v is an $\mathbb{R}^2 \times 1$ column vector, v^* is a $1 \times \mathbb{R}^2$ row vector.
2. And what is vv^*? *Hint*: an $\mathbb{R}^2 \times \mathbb{R}^2$ matrix.
3. Write it explicitly. *Hint*:
$$vv^* = v(x,p)\bar{v}(\tilde{x},\tilde{p}),$$
where (x,p) is the row index and $(\tilde{x},\tilde{p})$ is the column index.
4. This is the density matrix (at v).
5. Is it Hermitian?
6. Is it positive semidefinite?
7. What are its eigenvectors and eigenvalues? *Hint*: its nonsingular eigenvector is v itself (eigenvalue: 1). Indeed, thanks to associativity and normalization,
$$(vv^*)\, v = v\, (v^*v) = v\|v\|^2 = v.$$
Besides, vv^* also has many more singular eigenvectors. In fact, pick any vector orthogonal to v. This will be an eigenvector in its own right (eigenvalue: 0).
8. Look at vv^* again. What is its geometrical meaning? *Hint*: orthogonal projection onto v.
9. What is its trace? *Hint*:
$$\text{trace}\,(vv^*) = \int_{-\infty}^{\infty}\int_{-\infty}^{\infty} v(x,p)\bar{v}(x,p)dxdp = \|v\|^2 = 1.$$
10. Prove this in yet another way. *Hint*: look at v and v^* as (very thin) rectangular matrices. In the trace, you can reverse them and work the other way around:
$$\text{trace}\,(vv^*) = \text{trace}\,(v^*v) = v^*v = (v,v) = \|v\|^2. = 1$$

11. What is the expected energy at v? *Hint*:

$$v^*Hv = (v, Hv).$$

12. If H is already in its diagonal form, how to calculate this? *Hint*: in the diagonal form, X^2 acts horizontally and P^2 vertically.
13. Write the expected energy as a trace. *Hint*: in the trace, place v first or last — it doesn't matter:

$$\text{trace}\,(vv^*H) = \text{trace}\,(v^*Hv) = v^*Hv.$$

14. How to average vv^* with respect to x and drop both x and $\tilde{x}$ at the same time? *Hint*: integrate vv^* along the oblique line $x = \tilde{x}$, and obtain a new "smaller" $\mathbb{R} \times R$ matrix (named D_p), with the following elements:

$$D_p(p, \tilde{p}) \equiv \int_{-\infty}^{\infty} v(x, p)\bar{v}(x, \tilde{p})dx$$

(where p is the row index and $\tilde{p}$ is the column index).
15. What is this physically? *Hint*: this is the correlation between two wave functions: one at p and the other at $\tilde{p}$ (on two different horizontal lines).
16. D_p is the density matrix (for momentum alone, since x was already integrated upon and gone).
17. Is D_p Hermitian?
18. Is D_p positive semidefinite? *Hint*: D_p is actually an integral (or sum) of positive semidefinite matrices.
19. What does this tell you about the eigenvalues of D_p? *Hint*: they are positive (or zero).
20. In D_p, write the elements in yet another way. *Hint*:

$$D_p(p, \tilde{p}) = \int_{\rho(x)>0} \frac{v(x,p)}{\sqrt{\rho(x)}}\rho(x)\frac{\bar{v}(x,\tilde{p})}{\sqrt{\rho(x)}}dx$$

(where p is the row index and $\tilde{p}$ is the column index).
21. What is this? *Hint*: this is the expected density (correlation between p and $\tilde{p}$). Indeed, in the integrand, we multiplied by $\rho(x)$ (the probability to find the particle at x).

8.1.2 *Eigenvalues and Trace*

1. In a square (complex) matrix, sum the eigenvalues (with multiplicity). How to do this? *Hint*: if a few (linearly independent) vectors share the same eigenvalue, then add it a few times.
2. What do you get? *Hint*: this sum is the same as the trace (sum of main-diagonal elements).
3. Prove this. *Hint*: this follows from the Jordan form (see the following).
4. Look at the density matrix again: D_p (in its original form). What is its trace? *Hint*:

$$\begin{aligned}\text{trace}\,(D_p) &= \int_{-\infty}^{\infty}\int_{-\infty}^{\infty} v(x,p)\bar{v}(x,p)dxdp \\ &= \int_{-\infty}^{\infty}\int_{-\infty}^{\infty} |v|^2 dxdp = \|v\|^2 = 1.\end{aligned}$$

5. In D_p, integrate eigenvalue density (per unit momentum). How to do this? *Hint*: in the Jordan matrix, integrate along the main diagonal.
6. To make this possible, what physical unit should the eigenvalue density have? *Hint*: 1/momentum.
7. Why? *Hint*: this way, the eigenvalue is indeed a pure number, as required (with no unit at all).
8. What is the physical unit of the original main-diagonal density in D_p? *Hint*: 1/momentum too.
9. Why? *Hint*: because the v-density has unit $1/\sqrt{\text{length} \cdot \text{momentum}}$, and, to define D_p, vv^* was already integrated with respect to dx.
10. In terms of units, do these densities agree?
11. In the above integration (of eigenvalue density of D_p), what do you get? *Hint*: same as the trace: 1.
12. What does this tell you about the eigenvalues of D_p? *Hint*: they are between 0 and 1, and sum to 1.
13. Prove this. *Hint*: recall that D_p is Hermitian and positive semidefinite. Now, in the Jordan matrix, along the main diagonal, integrate eigenvalue density (per unit momentum), with respect to dp. This will sum the eigenvalues, as required (with their multiplicity).
14. Can D_p have an eigenvalue bigger than 1/2?

15. How many (linearly independent) eigenvectors can share it? In other words, what is the order of its Jordan block? *Hint*: at most one.
16. Why? *Hint*: otherwise, the trace would exceed 1, which is impossible for D_p.

8.2 Entanglement Entropy

8.2.1 *Entanglement Entropy*

1. Look at our density matrix again: D_p (for momentum alone, since x was already integrated upon, and gone). Define its entropy:

$$-\text{trace}\,(D_p \log D_p).$$

2. This is entanglement entropy.
3. What can you say about it? *Hint*: it is positive or zero.
4. Prove this. *Hint*: D_p has its eigenvalues between 0 and 1. To calculate its entropy, you can assume that D_p was already diagonalized (see the following).
5. Look at the entanglement entropy again. What does it tell us physically? *Hint*: the bigger it is, the more can information flow from x into p and even affect p and its (momentum) wave function and distribution.
6. This can even affect energy (indirectly). How? *Hint*: for simplicity, assume that H is already diagonal: the potential part acts horizontally and the kinetic vertically. Now, look at a physical process of two stages. In the beginning, H and P don't commute. Suddenly, they change and start to commute (so not only energy but also momentum is conserved). For example, the potential suddenly disappears, leaving $H = P^2/(2m)$ (kinetic energy only), expected to be

$$(v, Hv) = v^*Hv = \text{trace}\,(vv^*H) = \frac{1}{2m}\text{trace}\,\left(D_pP^2\right)$$

(where m is mass). In such a case, although x has no direct effect, it still affects energy indirectly (through p).
7. In the entanglement entropy, you need to calculate the logarithm of a matrix. How to do this? *Hint*: see the following.

8.2.2 *Matrix Logarithm*

1. Let's calculate the logarithm of a matrix. This will help calculate the entanglement entropy.
2. Let A be a square (complex) matrix. (This is most general. In our context, on the other hand, things will be even simpler: A will be Hermitian and positive semidefinite.)
3. Assume that $\|A\|$ is small enough. Why is this plausible? *Hint*: if $\|A\|$ is not small enough, then pick an integer m (Chapters 1 and 2 in Ref. [24]), and calculate
$$\log\left(2^m A/2^m\right) = m(\log 2)I + \log\left(A/2^m\right)$$
(where I is the identity matrix of the same order as A).
4. Show that
$$(I - A)^{-1} = I + A + A^2 + \cdots = \sum_{j=0}^{\infty} A^j.$$
Hint: multiply both sides by $I-A$. On the right-hand side, cancel opposite terms telescopically.
5. Integrate this, term by term. *Hint*:
$$-\log(I - A) = \sum_{j=1}^{\infty} \frac{A^j}{j}.$$
6. Substitute $I - A$ for A:
$$-\log A = \sum_{j=1}^{\infty} \frac{(I - A)^j}{j}.$$
7. Does this converge (in norm)? *Hint*: if the eigenvalues of A have a positive real part, then pick m so big (to make $\|A\|$ so small) that $I-A$ has its eigenvalues well inside the unit circle. This will guarantee convergence: in the above series, all terms commute and share the same eigenvectors, and even the same Jordan form (see the following).
8. Assume that A can be factorized as a triple product:
$$A = CBC^{-1},$$
for some matrices B and C. (Don't confuse B with the diagonal matrix in Section 1.5.1.)

9. How does this help calculate the trace of A? *Hint*:

$$\text{trace}(A) = \text{trace}\left(CBC^{-1}\right) = \text{trace}\left(C^{-1}CB\right) = \text{trace}(B).$$

10. In the above series, substitute CBC^{-1} for A, and pull C and C^{-1} out of parentheses:

$$\begin{aligned}
\log A &= -\sum_{j=1}^{\infty} \frac{(I-A)^j}{j} \\
&= -\sum_{j=1}^{\infty} \frac{(I-CBC^{-1})^j}{j} \\
&= -\sum_{j=1}^{\infty} \frac{(C(I-B)C^{-1})^j}{j} \\
&= -C\left(\sum_{j=1}^{\infty} \frac{(I-B)^j}{j}\right)C^{-1} \\
&= C\log(B)C^{-1}.
\end{aligned}$$

11. Conclude that, in the matrix logarithm, you can pull C and C^{-1} out of parentheses:

$$\log\left(CBC^{-1}\right) = C\log(B)C^{-1}.$$

12. Assume that B is diagonal (or at least upper-triangular):

$$B = \text{diag}\left(b_1, b_2, \ldots\right)$$

(plus some upper-triangular elements). How to calculate $\log B$ easily? *Hint*: in the above series for $\log A$, substitute B for A (on both sides) and work on the main diagonal on its own, element by element:

$$\log B = \text{diag}\left(\log b_1, \log b_2, \ldots\right)$$

(plus some upper-triangular elements).

13. In particular, this is true when $A = CBC^{-1}$ is the Jordan form of A (so A has eigenvalues b_1, b_2, $\ldots$).

14. Conclude that the eigenvalues of $\log(A)$ are the logarithm of the corresponding eigenvalues of A.
15. In other words, the logarithm is invariant (in terms of eigenvalues): the eigenvalue of the logarithm is the logarithm of the eigenvalue.
16. In our context, A will be Hermitian (and even positive semidefinite). Conclude that, in this case, C is unitary and B is diagonal (with $b_1 \geq 0$, $b_2 \geq 0$, ...).
17. In the following limit, let ε approach zero (from the right, so $\varepsilon > 0$), and prove that

$$\varepsilon \log \varepsilon \to_{\varepsilon \to 0} 0.$$

Hint: use L'Hopital's rule:

$$\begin{aligned}
\lim_{\varepsilon \to 0} \varepsilon \log \varepsilon &= \lim_{\varepsilon \to 0} \frac{\log \varepsilon}{\varepsilon^{-1}} \\
&= \lim_{\varepsilon \to 0} \frac{(\log \varepsilon)'}{(\varepsilon^{-1})'} \\
&= \lim_{\varepsilon \to 0} \frac{\varepsilon^{-1}}{-\varepsilon^{-2}} \\
&= -\lim_{\varepsilon \to 0} \varepsilon \\
&= 0.
\end{aligned}$$

18. Conclude that, if A has eigenvalue 0 or 1 only, then

$$A \log A = \left(CBC^{-1}\right)\left(C \log(B) C^{-1}\right) = CB \log(B) C^{-1} = (0)$$

(the zero matrix). *Hint*: this is so only in our context, where B is diagonal. Otherwise, this is not quite the zero matrix but some other trace-free (traceless) matrix.

8.2.3 *Pure State vs. Mixed State*

1. In particular, this is true for our original density matrix (at v):

$$vv^* \log(vv^*) = (0).$$

2. Conclude that vv^* has no entropy at all.
3. What does this tell us physically? *Hint*: in our original x-p system, the original (theoretical) state v is pure: not a mix (sum) of two (hypothetical) states (decided by flipping a coin or something like that). In other words, v already contains all information: no more information can flow in.
4. Is v always pure? *Hint*: in the whole x-p system, v is always pure. For p alone, on the other hand, v can also be mixed. Indeed, we already saw an example where information flows from x into p and even affects energy. When is v pure (even for p alone)? Only when D_p has no entropy at all (a separable state — see the following).
5. In general, how to calculate $A \log A$ more directly? *Hint*: in the above series for $\log A$, multiply both sides by A:

$$-A \log A = \sum_{j=1}^{\infty} \frac{A(I-A)^j}{j}.$$

6. How to calculate this numerically? *Hint*: use Taylor (or diagonal Pad'e) approximation (Chapters 1 and 2 in Ref. [24]).
7. Which method is better (more stable and accurate)?
8. Better yet, avoid calculating this series explicitly at all. Instead, take the trace, term by term. After all, in entanglement entropy, this is all we need. *Hint*: recall that the trace is linear. In the above series, work term by term:

$$-\text{trace}(A \log A) = \sum_{j=1}^{\infty} \frac{1}{j} \text{trace}\left(A(I-A)^j\right).$$

9. How to do this in C++? *Hint*: use your (sparse-)matrix class [24]. In your loop, work term by term. To save memory, store just one matrix: $A(I-A)^j$. How to calculate it? Look at the older matrix: $A(I-A)^{j-1}$. Update it: multiply by $I-A$ (either from the right or from the left – the trace doesn't care). Take the trace, and add up, term by term, in your loop.
10. How to use this to calculate the entanglement entropy? *Hint*: in the above, substitute D_p for A.

8.3 Entanglement vs. Disentanglement

8.3.1 *Two Particles*

1. So far, we looked at just one particle: x was its position and p its momentum. These didn't commute: they had a nonzero commutator. This is why our position–momentum wave function was only theoretical, not physical.
2. Next, look at two particles. Let y be some measurable of one particle and q some measurable of the other.
3. Can y and q be measured simultaneously? *Hint*: yes! Since they belong to different particles, they commute.
4. Let $v(y, q)$ be their joint wave function. Normalize it to have norm 1. *Hint*: divide v by its own norm. Once this is done,
$$\|v\|^2 = \int_{-\infty}^{\infty} \int_{-\infty}^{\infty} |v|^2 dy dq = 1,$$
as required.
5. Let y be a fixed parameter. Look at the $\mathbb{R} \times \mathbb{R}$ matrix
$$v(y, q)\bar{v}(y, \tilde{q})$$
(where q is the row index and $\tilde{q}$ is the column index).
6. Geometrically, what is this? *Hint*: an orthogonal projection onto the yth row in v (up to normalization).
7. In the theory of parallel universes, this will project on one particular universe (corresponding to one particular y).
8. Is it Hermitian?
9. Is it positive semidefinite?
10. Integrate on dy. What do you get? *Hint*: this will be D_q: the density matrix for q alone (since y was already integrated upon, and gone). Its $(q, \tilde{q})$th element is
$$D_q(q, \tilde{q}) \equiv \int_{-\infty}^{\infty} v(y, q)\bar{v}(y, \tilde{q}) dy$$
(where q is the row index and $\tilde{q}$ is the column index).
11. Is it Hermitian and positive semidefinite too? *Hint*: it is the integral (or sum) of Hermitian positive semidefinite matrices.
12. Is this an orthogonal projection too? *Hint*: only at disentanglement (see the following).

8.3.2 Disentanglement: Separability

1. Algebraically, what is a disentangled state? *Hint*: see the following.
2. Look at a separable state that can be written as a product of the form
$$v(y, q) = u(y)w(q),$$
where u and w are some (normalized) functions (of one variable each).
3. What does this tell you about v? *Hint*: v is pure (even for q alone, and even for y alone). Indeed, v already contains all information about y (in u), and also about q (in w), and needs no more information from y on q or from q on y.
4. Are u and w differentiable and (square-)integrable? *Hint*: yes (since v is).
5. Look at two fixed parameters: $y \neq \tilde{y}$. In v, they index two rows: $v(y, \cdot)$, and $v(\tilde{y}, \cdot)$ (as functions of q only). Are they proportional (linearly dependent)? *Hint*: yes (thanks to separability).
6. This kind of proportionality means (and indeed defines) disentanglement.
7. Does this work the other way around too? Assume now that we know nothing about separability yet. Instead, we know something else: disentanglement: in v, all rows are proportional. More precisely, for every pair of fixed $y \neq \tilde{y}$, $v(y, \cdot)$ and $v(\tilde{y}, \cdot)$ are linearly dependent (as functions of q only). How to design u and w and guarantee separability? *Hint*: pick some y_0 and q_0. Look at the y_0th row in v. (Thanks to their proportionality, the rows contain no zero elements, which are immaterial and can be dropped anyway.) Use this row to define w:
$$w(q) \equiv v(y_0, q)$$
(for all the q's). Next, use the same proportionality to define u too (from a column in v:
$$u(y) \equiv \frac{v(y, q_0)}{w(q_0)}$$
(for all the y's). Thanks to the same proportionality again,
$$\frac{v(y, q)}{w(q)} = \frac{u(y)}{w(q_0)}$$

(for all y's and q's), or indeed

$$v(y,q) = \frac{u(y)w(q)}{w(q_0)},$$

as required. Finally, normalize both u and w.

8. Conclude that disentanglement is equivalent to separability. This will be useful in the following.

8.3.3 *Separability: No Entanglement Entropy*

1. Look at our separable state again. Physically, does it represent disentanglement? In what sense? *Hint*: y and q don't affect (or interact with) one another at all.
2. How to see this mathematically? *Hint*: thanks to separability, there is no entanglement entropy (see the following).
3. Indeed, in our separable state, how does the density matrix D_q look like? *Hint*:

$$D_q = \int_{-\infty}^{\infty} u(y)w(q)\bar{u}(y)\bar{w}(\tilde{q})dy = \text{trace}\,(uu^*)\,ww^*$$
$$= \|u\|^2 ww^* = ww^*.$$

4. What does this mean geometrically? *Hint*: D_q is an orthogonal projection onto w.
5. What is its entropy? *Hint*: zero.
6. Is v pure (even for q alone)? *Hint*: yes. Thanks to separability, $D_q = ww^*$, which has no entropy at all.
7. What does this mean for v? *Hint*: in v, q is independent and doesn't care about y at all. Information about y can never affect q or its wave function or distribution.
8. And what about y? Is v pure for y alone as well? *Hint*: separability is symmetric. Thus, D_y has no entropy either.
9. What does this mean physically? *Hint*: complete disentanglement! Both y and q don't care about each other: they can never affect (the wave function or distribution of) one another.
10. And what about the other way around? Assume now that we know nothing about separability yet. Instead, we only know that D_q has no entropy at all. Does this imply separability? *Hint*: recall that the eigenvalues of D_q lie between 0 and 1, and sum

to 1 (with their multiplicity: Section 8.1.2). Here, thanks to our new assumption, we know even more: no entropy. Thus, D_q has just two eigenvalues: 0 (with many eigenvectors) and 1 (with one normal eigenvector – let's call it w). Thus, D_q must be an orthogonal projection onto w:

$$D_q = ww^*.$$

This leads to separability. Otherwise, there were at least two fixed $y \neq \tilde{y}$ for which $v(y, \cdot)$ and $v(\tilde{y}, \cdot)$ are linearly independent, yet D_q integrates both and projects on both.

11. Thus, separability is the same as no entanglement entropy, which is the same as v being pure (even for q alone and even for y alone).

8.3.4 *Entanglement: Positive Entanglement Entropy*

1. So far, we looked at a separable state. But this is quite rare and special. Let's move on to a more realistic state.
2. Assume no separability any more. This is entanglement: y and q do affect (and depend on) one another. (For example, both particles were prepared in the same way: Section 4.5.1.) This way, v can't be decomposed as a product any more.
3. What can you say about the entanglement entropy now? *Hint*: it is now positive.
4. Prove this. *Hint*: if it were zero, then we'd have separability again (as discussed above), in violation of our current assumption.
5. What does this mean physically? *Hint*: entanglement: y and q do affect each other and add more information to one another. They can't ignore each other any more: information about one affects the (wave function and distribution of) the other as well.

8.4 Parallel Universes: Can Entropy Split?

8.4.1 *Collapse: Is It Real?*

1. Let the above particles fly away from one another for a long time. Then, measure the second particle, and uncover its (deterministic) value: $q = q_0$. What happens to v? *Hint*: v "collapses" and

is substituted by its q_0th column:

$$v \leftarrow v(y, q_0)\,\delta(q - q_0)$$

(up to a constant coefficient — see the following).

2. What is this algebraically geometrically? *Hint*: an orthogonal projection onto the line $q = q_0$.
3. If you like, disregard $q \neq q_0$ (which is no longer possible anyway).
4. Normalize our new v. *Hint*: divide by a constant, to make sure that

$$\int_{-\infty}^{\infty} |v(y, q_0)|^2\, dy = 1.$$

5. This is called the "collapse" of the original (joint) wave function.
6. What happened physically? *Hint*: this is disentanglement (of the new y from the new q). For them, we have now a new separable v.
7. What happened to the entanglement entropy (the entropy of the new D_y)? *Hint*: it fell to zero.
8. What does this mean physically? *Hint*: less uncertainty, and more information. In fact, q already gave us all it can: it can add nothing more to our knowledge about y or its wave function or distribution.
9. Where did this new information come from? *Hint*: from observing and measuring q.
10. Is the "collapse" reversible? *Hint*: projection is irreversible.
11. What does this tell you physically? *Hint*: the "collapse" is nonphysical. In quantum mechanics, everything is reversible. Information can't jump for nothing. Entropy can't fall down for nothing.
12. Is there a more physical explanation? *Hint*: use not projection but entanglement (to the entire experimental device). After all, the process takes place in a bigger system, containing our experimental equipment as well. For example, in the bigger y-q system, vv^* has no entropy all along.
13. But, at entanglement, this explanation still uses an (irreversible) projection. Is there a yet better explanation? *Hint*: before measurement, the total entropy splits among many hypothetical universes. To pick the correct universe, there is a need to carry out an experiment (see the following).

8.4.2 *Parallel Universes and Entropy*

1. How to fix this? How to keep the process reversible?
2. For this, we must keep entropy constant (even for y on its own): the effect from q mustn't fall. How to guarantee this?
3. For this, avoid the "collapse" altogether. How? *Hint*: interpret q as the result in a measuring device for the second particle. No need to look at it. This way, all options remain open (each at its own probability). Each would lead to a different (hypothetical) universe (at some probability). In our universe, $q = q_0$. This way, we have more information and less entropy. But, in other (parallel) universes, $q \neq q_0$. The total entropy remains the same: it only splits now among multiple universes. More precisely, to obtain D_y, integrate over all these universes.
4. Is this reversible? Can you get the genie back into the bottle? *Hint*: in theory, you can rewind the movie, back in time! This way, you'll make all these (parallel) universes shrink back into one. But, to see this, you must sit outside all universes. Sit, and have fun!
5. Still, at some point, we look at our measuring device and read the result: $q = q_0$. But this measurement is made at the second particle, far away from the first one (which is still here, in our lab). How could information about q arrive to us instantaneously (faster than light) and affect y (and its wave function and distribution)? This is the EPR paradox (Einstein–Podolsky–Rosen). *Hint*: from the start, we used our own logic (implicitly): the particles were prepared in the same way, so they must be the same. But this is not quite physical: they are only "entangled" logically and indirectly (through the equipment that produced them in the same way). We only called them "entangled" to help design v, which helped information "arrive" and entanglement entropy "fall." Thanks to our logic, we even used our background in conservation laws and deduced that there was no change over time. (See more in Section 4.5.1.)

8.5 Bell's Experiment

8.5.1 *Bell's Experiment*

1. In Bell's experiment, on the other hand, the particles are produced not in the same way but completely opposite to one another.

2. They are still entangled (and correlated) but in a new way: the correlation is now negative not positive any more.
3. This is called Bell's state. In it, the particles have not the same spin but the opposite spin (in every direction).
4. Let's accept this for now. (To prove this, some calculations are needed.)
5. Bell's state is now used in Bell's experiment, which contains two parts (Figure 8.1).
6. First, measure their spin in two perpendicular directions: say, $y = \pm 1$ is spin-up and $q = \pm 1$ is spin-right. This is the first part.
7. In the second part, on the other hand, pick new directions: no longer perpendicular but at 45° from one another.
8. Do this in two versions.
9. In the first version, y is still spin-up, but q is now spin-up-right.
10. In the second version, on the other hand, y is spin-up-right, and q is spin-right. This completes the whole experiment.
11. It turns out that the second part is less likely: add the probability of both versions, and you'll still get less than the probability of the first part! How come? *Hint*: in the second part, the system is actually bigger: it also contains a new (oblique) measuring device,

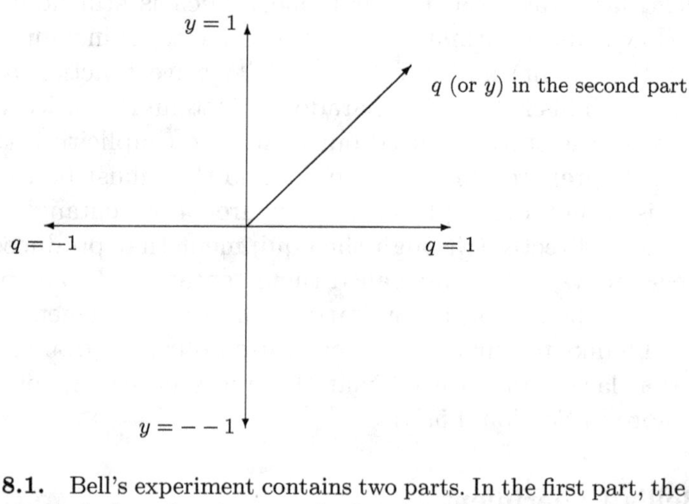

Figure 8.1. Bell's experiment contains two parts. In the first part, the particles spin perpendicularly: y is spin-up and q is spin-right. In the second part, on the other hand, they spin at 45° from one another. This is done in two versions. In the first version, y is still spin-up, but q is now spin-up-right. In the second version, on the other hand, y is spin-up-right and q is spin-right.

in between. As a matter of fact, the system is even bigger: the first version changed the original state completely. To have it back, the second version must prepare a new pair of particles, in the same way as before.

12. Why is nothing faster than light? *Hint*: many virtual particles stand in the way (even in vacuum).

8.6 Double-Slit: Axiom of Choice

8.6.1 *Double-Slit: Axiom of Choice*

1. In the double-slit experiment (Figure 4.1), we have two (horizontal) slits: upper and lower. Index the upper by 1 and the lower by 0.
2. Define a new (binary) random variable:

$$b \equiv \begin{cases} 1 & \text{if the particle went through the upper slit} \\ 0 & \text{if the particle went through the lower slit.} \end{cases}$$

3. First, assume that b is never recorded.
4. Is there any interference on the screen behind? *Hint*: yes.
5. Next, assume that b is recorded and saved (before the particle hits the screen on the right).
6. What is this physically? *Hint*: measurement.
7. Is there any interference on the screen behind? *Hint*: not any more.
8. Can this be viewed as entanglement with some measuring device? *Hint*: to record which slit was picked, design a new spin that stores b. Initially, this new spin is set to 0 (up). Then, it turns into 1 (down), provided that the particle went through the upper slit (before hitting the screen on the right). Otherwise, it remains 0.
9. What is this physically? *Hint*: measurement.
10. What is this mathematically? *Hint*: indexing: b is the index of the slit that was picked.
11. How to change things and have interference back again? *Hint*: b must remain implicit and nondeterministic. For example, assume that the slits are indistinguishable (so we can never tell which is which). This way, even if we know b, we gain no more information about the particle or its history.

12. Next, make more slits: in the screen on the right (where the interference pattern was displayed), cut two new (horizontal) slits, to let the particle go through (rightward: Figure 8.2).
13. Behind this, to the right, place yet another screen, with two slits in it as well, and so on, screen by screen (infinitely many screens, to the far right).
14. To travel rightward (to $+\infty$), the particle must now go through infinitely many screens (screen by screen), never telling us through which slit. Can it do this? *Hint*: thanks to the axiom of choice, there is a choice function

$$c: \text{ all screens} \to \{0, 1\},$$

with the following physical meaning:

$$c(\text{screen}) = \begin{cases} 1 & \text{if the upper slit is picked at this screen} \\ 0 & \text{if the lower slit is picked at this screen.} \end{cases}$$

15. What does this mean physically? *Hint*: the particle can indeed go through and arrive safely to $+\infty$. Still, c remains implicit and is never recorded. (If you like, the slits remain indistinguishable.) This leads to interference at $+\infty$.

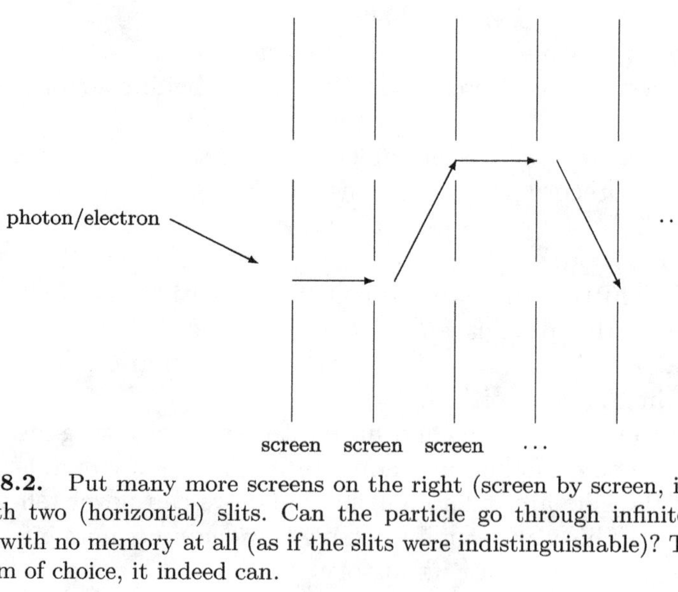

Figure 8.2. Put many more screens on the right (screen by screen, in a row), each with two (horizontal) slits. Can the particle go through infinitely many screens, with no memory at all (as if the slits were indistinguishable)? Thanks to the axiom of choice, it indeed can.

16. This is probably how nature really works.
17. Compare this to Chapter 12 in Ref. [27]. Are these proofs equivalent? In what sense?
18. Download a random-number generator from the Internet, and use it to simulate spin on your digital computer.
19. Use this to simulate a quantum computer.
20. Is there a real quantum computer? *Hint*: not yet.
21. Use your code to implement quantum FFT. The solution follows from Chapters 14 and 18.
22. Use quantum FFT to implement Shor's algorithm (to break the RSA key exchange in cryptography). The solution follows from Chapter 15.

Part III: Angular Momentum and Spin

Like energy, angular momentum takes only discrete values. Unlike energy, however, it takes only a few positive values and also their negative counterparts. This is quantum mechanics for you: both energy and angular momentum come in quanta, or packets.

Spin is the most elementary form of angular momentum. It takes only two possible values: a positive value and its negative counterpart. This is indeed an elementary random variable, with only two options: either spin-up (couterclockwise cycling) or spin-down (clockwise cycling).

This will be the basis for quantum computing. Nowadays, no quantum computer is available yet. Still, we can imagine one and write interesting quantum algorithms for it. We'll come back to this later.

Chapter 9

Energy and Angular Momentum: An Algebraic Study

How to study energy and angular momentum at the same time? For this, we develop a uniform framework, based on linear algebra. Indeed, thanks to the commutator, we manage to design ladder operators, to climb up or down the ladder, and obtain many energy levels, on which energy is conserved. Still, they are discrete, not continuous. This is quantum mechanics for you: energy comes in packets, or quanta.

Likewise, angular momentum can also benefit from the same methodology. Still, it is fundamentally different. Unlike energy, which has infinitely many (positive) levels, angular momentum has only a few eigenvalues: positive and negative alike, mirroring each other.

Energy is unbounded. In theory, it can take very high values, as high as you like. In practice, on the other hand, high energy is highly unlikely, uncertain, and even unstable: it can even leak out and fall down, back to low energy, which is much more likely, steady, and attractive. We already used this principle in AI too.

Angular momentum, on the other hand, is strictly bounded: it takes only a few values (positive and negative alike). The most elementary example is spin. It takes only two possible values: a positive one, mirrored by a negative one. Later on, we'll see how spin is related to another interesting phenomenon: the polarization of light. Moreover, spin will be the key to a new field: quantum computing and algorithms.

9.1 Heisenberg's Uncertainty Principle

9.1.1 *The Cauchy–Schwarz Inequality*

So far, we already used Heisenberg's uncertainty principle in a few places. Still, how to prove it algebraically? For this, we need some preliminaries from linear algebra. Look at two wave functions: u and v:

$$u, v : \mathbb{R} \to \mathbb{C}.$$

Look at their inner product: $(u, v) \in \mathbb{C}$. How to estimate it (in absolute value)? For this, use their norms: $\|u\|$ and $\|v\|$:

$$|(u, v)| \leq \|u\| \cdot \|v\|.$$

This is the Cauchy–Schwarz inequality. Let's go ahead and use it.

9.1.2 *Hermitian and Anti-Hermitian Parts*

Next, take our position operator X and our momentum operator P. They are both Hermitian. Now, multiply them by one another. This will give you a new operator: the product XP. But this is not Hermitian any more. Split it into two parts. In other words, write it as a sum of Hermitian part, plus an anti-Hermitian part:

$$\begin{aligned} XP &= \frac{XP + (XP)^*}{2} + \frac{XP - (XP)^*}{2} \\ &= \frac{XP + PX}{2} + \frac{XP - PX}{2} \\ &= \frac{XP + PX}{2} + \frac{[X, P]}{2}. \end{aligned}$$

Next, use this to calculate the expectation of XP at u:

$$(u, XPu) = \left(u, \frac{XP + PX}{2} u\right) + \left(u, \frac{[X, P]}{2} u\right).$$

What do we have here? On the left-hand side, we have a complex number. On the right-hand side, it splits into its real part, plus its imaginary part. Clearly, each part is smaller than (or equal to) the original complex number (in absolute value). In particular,

$$|(u, XPu)| \geq \frac{1}{2}|(u, [X, P]u)| = \frac{1}{2}|(u, i\bar{h}u)| = \frac{\bar{h}}{2}(u, u) = \frac{\bar{h}}{2}$$

(assuming that u was already normalized). Let's go ahead and use this.

9.1.3 *Standard Deviation*

For simplicity, assume that we already shifted X. In other words, we already subtracted a constant number: (u, Xu) (the expected position at u). This way, our up-to-date X already has expectation zero at u. Likewise, assume that we also did the same for P as well. This will simplify things considerably.

Now, what is the standard deviation of X at u? This is $\|Xu\|$. It tells us about the values that X could take and how spread-out they are. Now, take the standard deviation of X at u, times the standard deviation of P at u. How small can this be? How to estimate this from below? Thanks to the Cauchy–Schwarz inequality (and all what we did above),

$$\|Xu\| \cdot \|Pu\| \geq |(Xu, Pu)| = |(u, XPu)| \geq \frac{\bar{h}}{2}.$$

This is Heisenberg's uncertainty principle: the key to quantum mechanics. It tells us that, at the original state u, X and P can't be measured simultaneously. Indeed, if X was measured exactly, then its standard deviation would fall to zero at u. The above inequality, would then force P to have an infinite standard deviation, which means that P could never be measured at u any more. The same is true the other way around: if P was measured exactly, then X could never be measured any more.

9.2 Commutator and Eigenvalues

9.2.1 *Shifting an Eigenvalue*

How to study both energy and angular momentum at the same time? For this, look at the commutator in general. Let C and T be two operators (Hermitian or not). Assume that they don't commute but have a nonzero commutator, proportional to T:

$$[C, T] \equiv CT - TC = \alpha T,$$

for some (complex) parameter $\alpha \neq 0$. (If $\alpha = 0$, then the following is still true but less interesting.) Let u be an eigenvector of C, with eigenvalue λ:

$$Cu = \lambda u.$$

How to find more eigenvectors? Just apply T to u. Indeed, if $Tu \neq \mathbf{0}$, then this is an eigenvector in its own right:

$$C(Tu) = (CT)u = (TC + [C, T])\, u = (TC + \alpha T)\, u = (\lambda + \alpha)\, Tu.$$

Thus, we managed to shift λ by α and obtain a new eigenvalue of C: $\lambda + \alpha$ (so long as $Tu \neq \mathbf{0}$).

Likewise, apply T time and again (so long as you don't hit the zero vector), and obtain more and more eigenvalues of C:

$$\lambda, \lambda + \alpha, \lambda + 2\alpha, \lambda + 3\alpha, \ldots.$$

This could go on and on forever.

9.2.2 *Shifting an Eigenvalue of a Product*

Let's use this in a special case: a product. Let A and B be two operators (Hermitian or not). Assume that they don't commute but have a nonzero commutator, proportional to the identity operator:

$$[A, B] \equiv AB - BA = \beta I,$$

for some (complex) parameter $\beta \neq 0$. (If $\beta = 0$, then the following is still true, but less interesting.) Now, look at their product: BA. What is its commutator with B? We already know what it is:

$$[BA, B] \equiv BAB - BBA = B(AB - BA) = B[A, B] = \beta B.$$

So, in Section 9.2.1, we can now substitute

$$C \leftarrow BA, \quad \text{and} \quad T \leftarrow B.$$

This will help shift an eigenvalue of BA. For this, let u be an eigenvector of BA, with eigenvalue λ:

$$BAu = \lambda u.$$

If $Bu \neq \mathbf{0}$, then this is an eigenvector of BA in its own right:

$$BA(Bu) = (\lambda + \beta)Bu.$$

How to shift in the opposite direction? In Section 9.2.1, substitute

$$C \leftarrow BA, \quad \text{and} \quad T \leftarrow A.$$

Indeed, their commutator is

$$[BA, A] = BAA - ABA = (BA - AB)A = -[A, B]A = -\beta A.$$

Thanks to this, we can now design a new eigenvector of BA: take the original eigenvector u, and apply A to it:

$$BA(Au) = (\lambda - \beta)Au.$$

If $Au \neq \mathbf{0}$, then this is an eigenvector of BA in its own right, with a new eigenvalue: $\lambda - \beta$. Likewise, go on and on (so long as you don't hit the zero vector), and obtain more and more eigenvalues of BA:

$$\lambda, \lambda \pm \beta, \lambda \pm 2\beta, \lambda \pm 3\beta, \ldots.$$

This could go on and on forever.

9.2.3 *A Number Operator*

Next, look at an interesting special case:

$$B \equiv A^* \quad \text{and} \quad \beta \equiv 1.$$

This way, the assumption in the beginning of Section 9.2.2 takes the form:

$$[A, A^*] = I.$$

Furthermore, our product is now

$$BA = A^*A.$$

Since A^*A is Hermitian and positive semidefinite, its eigenvalues are positive (or zero). Furthermore, thanks to the above discussion, we can now lower or raise an eigenvalue of A^*A: from λ to

$$\lambda + 1, \lambda + 2, \lambda + 3, \ldots$$

(until hitting the zero vector) and also to

$$\lambda - 1, \lambda - 2, \lambda - 3, \ldots$$

(until hitting the zero vector, which must be very soon).

A^*A is called a number operator. Why? Because its eigenvalues are the natural numbers: $0, 1, 2, 3, 4, \ldots$. Indeed, as we'll see soon, the raising process never stops.

9.2.4 *Eigenvalue: Expectation*

Let u be an eigenvector of A^*A, with eigenvalue λ:

$$A^*Au = \lambda u.$$

Since A^*A is Hermitian and positive semidefinite, λ must be positive (or zero). Indeed, A^*A must have a positive (or zero) expectation at u:

$$\lambda(u, u) = (u, A^*Au) = (Au, Au) \geq 0.$$

Let's use u to design more eigenvectors. For a start, let's lower.

9.2.5 *Down the Ladder*

By now, in Section 9.2.2, we already set $B \equiv A^*$, and $\beta \equiv 1$. This way, from our original eigenvector u, we can now design a new eigenvector: Au (if $Au \neq \mathbf{0}$), with a lower eigenvalue: $\lambda - 1$. Thus, A serves as a ladder operator: by applying A to u, we go down the ladder, to a yet smaller eigenvalue. This is why A is often denoted by

$$A^- \equiv A.$$

What is the norm of Au? To find out, look again at the expectation of A^*A at u:

$$\|Au\|^2 = (Au, Au) = (u, A^*Au) = \lambda(u, u) = \lambda\|u\|^2.$$

In other words,

$$\|Au\| = \sqrt{\lambda}\|u\|.$$

Later on, we'll use this to normalize the eigenvectors.

9.2.6 *Null Space*

This lowering process can't go on forever, or we'd eventually hit a negative eigenvalue, which is forbidden (Section 9.2.4). It must terminate upon hitting some eigenvector w in the null space of A^*A:

$$A^*Aw = \mathbf{0}.$$

At this point, lowering must stop. Thus, we can deduce in retrospect that λ must have been a nonnegative integer number.

So far, we saw that w is in the null space of A^*A. Is w in the null space of A too? In other words, is $Aw = \mathbf{0}$? Well, at w, A^*A must have zero expectation, and $\|Aw\|$ must vanish:

$$(Aw, Aw) = (w, A^*Aw) = (w, \mathbf{0}) = 0.$$

As a result,

$$Aw = \mathbf{0}.$$

So, w is in the null space of A too. Starting from w, we can't lower any more. Still, we can now work the other way around: raise back again.

9.2.7 *Up the Ladder*

How to raise eigenvalues up again? Use Section 9.2.2 once again (with $B \equiv A^*$, and $\beta \equiv 1$, as before). This time, look at A^*u: a new eigenvector of A^*A, with a bigger eigenvalue: $\lambda + 1$. Here, A^* serves as a new ladder operator, to "climb" up. This is why A^* is often denoted by

$$A^+ \equiv A^*.$$

What is the norm of A^*u? Well, since $[A, A^*] = I$, we have

$$\begin{aligned}
\|A^*u\|^2 &= (A^*u, A^*u) \\
&= (u, AA^*u) \\
&= (u, (A^*A + [A, A^*])\, u) \\
&= (u, (A^*A + I)\, u) \\
&= (u, (\lambda + 1)\, u) \\
&= (\lambda + 1)\, (u, u)\, .
\end{aligned}$$

In other words,

$$\|A^*u\| = \sqrt{\lambda + 1}\|u\|.$$

This is why the raising process can never stop. Later on, we'll use this formula to normalize these eigenvectors, as required.

9.3 Example: Harmonic Oscillator

9.3.1 *Ladder Operators*

Actually, we already saw an example: the harmonic oscillator (Section 2.4). To raise an eigenvalue, we already defined

$$A^+ \equiv \sqrt{\frac{m\omega}{2\bar{h}}}\left(X - \frac{i}{m\omega}P\right)$$

(Section 3.1.2). To lower an eigenvalue, on the other hand, use its Hermitian adjoint:

$$A^- \equiv (A^+)^* = \sqrt{\frac{m\omega}{2\bar{h}}}\left(X + \frac{i}{m\omega}P\right)$$

(since X and P are both Hermitian). This way, we indeed have

$$\begin{aligned}
[A^-, A^+] &= \frac{m\omega}{2\bar{h}}\left[X + \frac{i}{m\omega}P, X - \frac{i}{m\omega}P\right] \\
&= -\frac{m\omega}{2\bar{h}} \cdot \frac{i}{m\omega}\left([X, P] - [P, X]\right) \\
&= -\frac{i}{2\bar{h}} 2i\bar{h}I \\
&= I,
\end{aligned}$$

as required. Thus, A^+A^- is a legitimate number operator. Besides, it takes part in the original factorization of the Hamiltonian:

$$H = \omega\bar{h}\left(A^+A^- + \frac{1}{2}\right)$$

(Section 2.4.2). This is how we got our good old energy levels:

$$\frac{\omega\bar{h}}{2}, \frac{3\omega\bar{h}}{2}, \frac{5\omega\bar{h}}{2}, \ldots.$$

The first one is vacuum energy, corresponding to the ground state. How does it look like?

9.3.2 *Ground State*

In our harmonic oscillator, what is w? Well, it solves

$$A^- w = 0.$$

This is the ground state, of minimal energy: vacuum energy. It looks like a real Gaussian (Figure 9.1). Apply A^+ to it, time and again (on the x-axis), and obtain more and more eigenfunctions (of both the Hamiltonian and the number operator), of higher and higher energy, conserved all the time (Section 3.1.3). Better yet, on your way up the ladder, normalize your eigenfunctions, to look like this:

$$w, A^+ w, \frac{1}{\sqrt{2!}} \left(A^+\right)^2 w, \frac{1}{\sqrt{3!}} \left(A^+\right)^3 w, \ldots .$$

The same kind of normalization was already used in the Fourier series (Section 2.5.1).

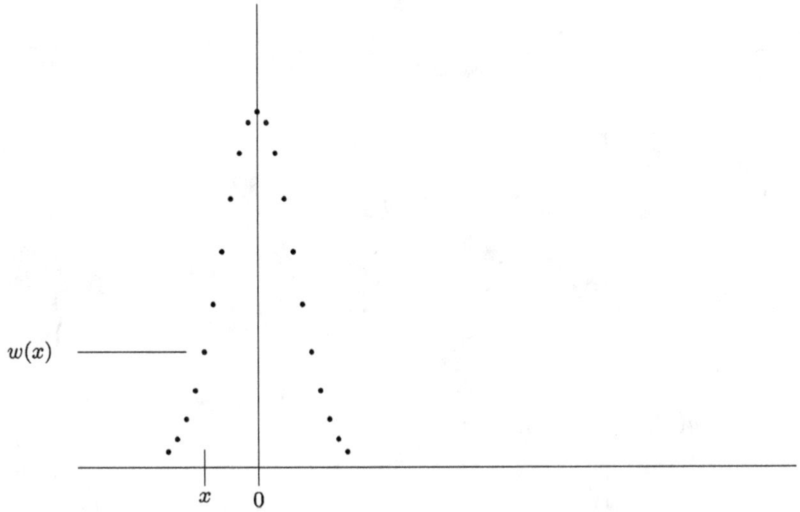

Figure 9.1. The ground state is the (real) Gaussian. This is w (in the null space of both A^- and A^+A^-, with zero expectation). To be at x, the particle has probability $|w(x)|^2$ (normalized). This is why it is highly likely to be at $x = 0$ (the expected position).

9.3.3 *Coherent State*

The coherent state, on the other hand, is more general: it solves

$$A^{-}u = \lambda u,$$

with eigenvalue

$$\lambda = \sqrt{\frac{m\omega}{2\hbar}}\mu.$$

How does u look like? It looks like a complex Gaussian (centered at μ: Figure 9.2).

9.3.4 *Energy: Poisson Distribution*

Once normalized properly, $|w|^2$ (or $|u|^2$) tell us how position is distributed over the entire x-axis. And how about energy? How is it distributed over infinitely many energy levels? This is the (discrete) Poisson distribution (Figure 9.3 and Section 2.5.3). As you can see, high energy is highly unlikely. Moderate energy, on the other hand, is highly likely, stable, and robust. We already used this in AI.

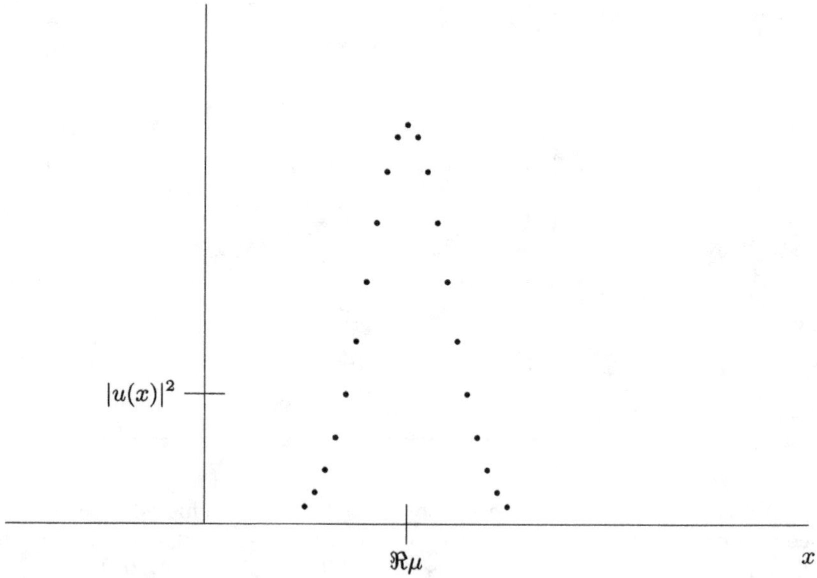

Figure 9.2. The coherent state looks like a complex Gaussian, centered at μ. To be at x, the particle has probability $|u(x)|^2$ (normalized).

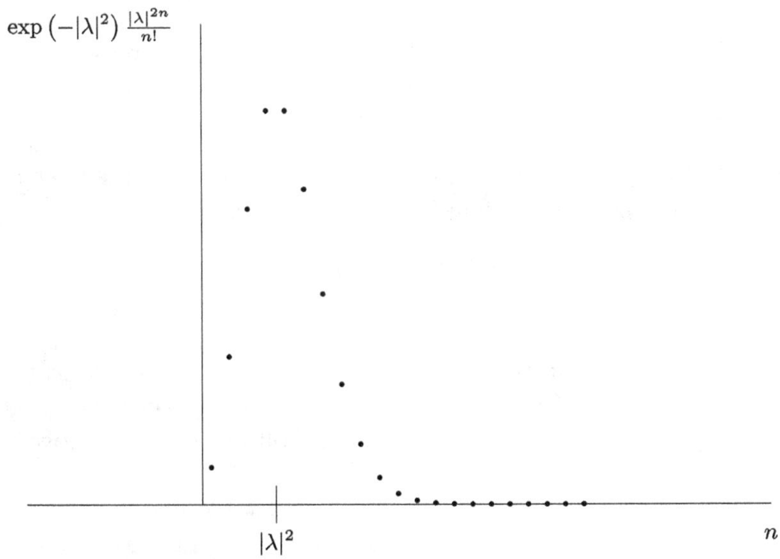

Figure 9.3. The Poisson distribution. To have energy $\omega\bar{h}(n + 1/2)$, the probability is $\exp(-|\lambda|^2)|\lambda|^{2n}/n!$. The maximum is at $n \doteq |\lambda|^2$.

9.4 Particle in 3-D

9.4.1 *Energy vs. Angular Momentum*

How is energy different from angular momentum? Energy has infinitely many (positive) levels. Angular momentum, on the other hand, has only a few: positive ones, mirrored by their negative counterparts. Why? Because energy is unbounded: in theory, it can reach very high levels, as high as you like. In practice, however, they are not very useful: they are highly unlikely, unstable, and uncertain. Low levels, on the other hand, are much more stable and attractive. This is why they are used in AI.

Angular momentum, on the other hand, is bounded. This is why it takes only a few possible values: half of them positive and half negative. Later on, we'll see the ultimate example: spin, useful in quantum computing and algorithms.

9.4.2 *Wave Function*

To introduce angular momentum (in its quantum-mechanical face), we need to move to 3-D. So far, our particle was in 1-D: confined to

$x \in \mathbb{R}$. This was quite theoretical and easy to model. Let's move on to reality: 3-D. Our new wave function v is now defined in 3-D:

$$v : \mathbb{R}^3 \to \mathbb{C}.$$

This new v will tell us about probability. For this, assume that v was already normalized (so $|v|^2$ has integral 1 in $\mathbb{R}^3$):

$$\|v\|^2 = (v, v) = \int_{-\infty}^{\infty} \int_{-\infty}^{\infty} \int_{-\infty}^{\infty} |v|^2 dxdydz = 1.$$

Now, how likely is the particle to be at a particular point $(x, y, z) \in \mathbb{R}^3$? The probability for this is $|v(x, y, z)|^2$. More precisely, $|v|^2$ is the probability density. To use it, consider a volume $V \subset \mathbb{R}^3$. How likely is the particle to be in V? The probability for this is an integral over V:

$$\iiint_V |v|^2 dxdydz.$$

Let's define new operators that can act on v.

9.4.3 *Position Operators*

How to define position operators in 3-D (to act on functions like v)? X will now be x-position (regardless of y or z). Likewise, define Y and Z too:

$$X = x\cdot, \quad Y = y\cdot, \quad \text{and} \quad Z = z\cdot.$$

Clearly, they commute. Let's define momentum operators too.

9.4.4 *Momentum Operators*

How to define momentum operators? Well, P_x will be x-momentum (regardless of y or z, just like our good old 1-D P). Likewise, define P_y (y-momentum) and P_z (z-momentum) as well:

$$P_x \equiv -i\bar{h}\frac{\partial}{\partial x}, P_y \equiv -i\bar{h}\frac{\partial}{\partial y}, \quad \text{and} \quad P_z \equiv -i\bar{h}\frac{\partial}{\partial z}.$$

9.4.5 *Commutator*

Clearly, they commute. Only three pairs don't commute:

$$[X, P_x] = [Y, P_y] = [Z, P_z] = i\bar{h}I$$

(where I is the identity operator in 3-D). How to write this more compactly? Let C and T stand for X or Y or Z. Then,

$$[C, T] = [P_c, P_t] = 0$$

(the zero operator in 3-D)

$$\text{and} \quad [C, P_t] = \begin{cases} i\bar{h}I & \text{if} \quad C = T \\ 0 & \text{if} \quad C \neq T. \end{cases}$$

Let's go ahead and use this.

9.5 Angular Momentum

9.5.1 *Angular Momentum Component*

In quantum mechanics, angular momentum will be nondeterministic: a random variable (or a measurable or an observable). This will mirror the good old (deterministic) angular momentum in classical mechanics. It will contain three (3-D) operators:

$$\begin{aligned} L_x &\equiv YP_z - ZP_y, \\ L_y &\equiv ZP_x - XP_z, \\ L_z &\equiv XP_y - YP_x. \end{aligned}$$

This is cyclic: the right-hand x-y-z coordinates shift cyclically. Now look at these operators. Are they Hermitian? Thanks to commutativity, they sure are. For instance,

$$L_x^* = (YP_z - ZP_y)^* = P_z^*Y^* - P_y^*Z^* = P_zY - P_yZ = YP_z - ZP_y = L_x$$

(thanks to Section 9.4.5).

9.5.2 *The Commutator*

By now, we already have three new operators: our new angular-momentum components. Do they commute? No! Indeed, they have a nonzero commutator:

$$\begin{aligned}[L_x, L_y] &= [YP_z - ZP_y, ZP_x - XP_z] \\ &= [YP_z, ZP_x] - [YP_z, XP_z] - [ZP_y, ZP_x] + [ZP_y, XP_z] \\ &= YP_x\,[P_z, Z] - YX\,[P_z, P_z] - P_yP_x\,[Z, Z] + P_yX\,[Z, P_z] \\ &= -i\bar{h}YP_x - 0 - 0 + i\bar{h}XP_y \\ &= i\bar{h}L_z.\end{aligned}$$

9.5.3 *An Uncertainty Principle*

What does this mean physically? They obey an uncertainty principle of their own: if L_x was measured exactly, then L_y can never be measured any more on the original state (and vice versa). Now, the above relation extends cyclically:

$$\begin{aligned}[L_x, L_y] &= i\bar{h}L_z, \\ [L_y, L_z] &= i\bar{h}L_x, \\ [L_z, L_x] &= i\bar{h}L_y.\end{aligned}$$

What does this mean physically? If one angular-momentum component was measured exactly, then the other two can never be measured (on the original state) any more. The above relations reflect this algebraically. Later on, we'll mirror this in Pauli matrices. Let's go ahead and use this.

9.5.4 *Up the Ladder*

Let's use this to raise an eigenvalue. For example, let u be an eigenvector of L_z, with eigenvalue λ:

$$L_z u = \lambda u.$$

(Since L_z is Hermitian, λ must be real.) How to raise λ? Define a new operator:

$$T \equiv L_x + iL_y.$$

This way, T is not Hermitian and doesn't commute with L_z. Indeed, they have a nonzero commutator:

$$\begin{aligned}[L_z, T] &= [L_z, L_x + iL_y] \\ &= [L_z, L_x] + i[L_z, L_y] \\ &= i\hbar L_y + \hbar L_x \\ &= \hbar T.\end{aligned}$$

Thanks to Section 9.2.1, we can now raise λ: if $Tu \neq \mathbf{0}$, then it is an eigenvector of L_z in its own right, with eigenvalue $\lambda + \hbar$. This way, T serves as a new ladder operator, to help "climb" up. This is why T is often denoted by

$$A^+ \equiv T.$$

This can be done time and again, producing bigger and bigger eigenvalues, until hitting the zero vector, and reaching the maximal eigenvalue of L_z. This is quantum mechanics for you: angular momentum is not continuous but can take only a few discrete values.

9.5.5 *Down the Ladder*

This also works the other way around: to not only raise an eigenvalue but also lower it. For this purpose, redefine T (this time with a minus sign):

$$T \equiv L_x - iL_y.$$

(This new T is the Hermitian adjoint of the old T.) As before, our new T doesn't commute with L_z. Indeed, they have a nonzero commutator again:

$$\begin{aligned}[L_z, T] &= [L_z, L_x - iL_y] \\ &= [L_z, L_x] - i[L_z, L_y] \\ &= i\hbar L_y - \hbar L_x \\ &= -\hbar T.\end{aligned}$$

Thanks to Section 9.2.1, we can now lower λ: if $Tu \neq \mathbf{0}$, then it is an eigenvector of L_z in its own right, with eigenvalue $\lambda - \hbar$. This way, our new T serves as a new ladder operator, to help go down the ladder. This is why our new T is often denoted by

$$A^- \equiv T.$$

Furthermore, this can be done time and again, producing lower and lower eigenvalues, until hitting the zero vector, and reaching the minimal eigenvalue of L_z.

So far, we worked with L_z. But there is nothing special about it: the same can be done for L_x and L_y as well. So, we have three operators, ready to serve as components in our new (quantum-mechanical) angular momentum.

9.5.6 *Angular Momentum*

Let's plug these operators as blocks in a new (rectangular) operator:

$$L \equiv \begin{pmatrix} L_x \\ L_y \\ L_z \end{pmatrix}.$$

This is our new (nondeterministic) angular momentum. It mirrors the good old (deterministic) angular momentum in classical mechanics. Physically, however, it is different: it obeys an uncertainty principle: only one angular-momentum component can be measured exactly but not the other two. In the following chapter, we'll see an interesting example: spin (with a lot of exercises too).

Chapter 10

Spin and Pauli Matrices

In quantum mechanics, both energy and angular momentum are discrete: they come in packets, or quanta. In fact, angular momentum takes only a few positive values and their negative counterparts. The most elementary example is spin. It takes just two values: either spin-up (counterclockwise) or spin-down (clockwise). The former is marked by $1/2$ and the latter by $-1/2$.

In our 3-D world, there are three spins: spin-up-or-down, spin-right-or-left, and spin-in-or-out. Each is nondeterministic: an uncertain random variable. For example, spin-up-or-down has a (discrete) wave function $v \equiv (v_0, v_1)^t$, telling us the probability $|v_0|^2$ to spin up or $|v_1|^2$ to spin down. More precisely, $v \equiv v(x)$ is also a function of position: x. Or, if you like, $v \equiv v(p)$ also depends on momentum: p. But here, we disregard this, for simplicity.

So, we have three (nondeterministic) angular-momentum components: spin up-or-down, right-or-left, and in-or-out. They are represented by operators that don't commute. Thus, only one of them can be measured exactly but not the other two. This is the uncertainty principle.

How to model these three spins? For this, we have an algebraic tool: Pauli matrices. Like our spins, they don't commute. On the contrary: they even anti-commute. Thanks to this property, they can even be used to design yet more advanced matrices: Dirac's matrices. These are useful to model electrons and positrons alike (and, in general, matter and anti-matter in particle physics). As discussed in Chapters 3 and 4, this is mirrored by $\mathbb{Z}$: the (additive) group of integer numbers.

10.1 Hamiltonian and Energy Levels

10.1.1 *Eigenvalues and Eigenvectors*

1. What is an observable? *Hint*: an observable is the same as measurable: a Hermitian matrix (or operator).
2. What can you say about its eigenvalues? *Hint*: they are real (no imaginary part).
3. Consider two eigenvectors, corresponding to two distinct eigenvalues. What can you say about them? *Hint*: they are orthogonal to each other.
4. Can you make them orthonormal? *Hint*: normalize them (to have norm 1).
5. And what if they share the same eigenvalue? Assume they are linearly independent. How to orthogonalize them? *Hint*: pick one of them, and modify it: subtract its orthogonal projection on the other one (obtained from their inner product). This is the Gram–Schmidt process.
6. Normalize them to have norm 1.
7. How many (linearly independent) eigenvectors does an observable have? *Hint*: as many as its order: n.
8. How many (distinct) eigenvalues can an observable have? *Hint*: at least one, and up to n.
9. A degenerate observable has less than n (distinct) eigenvalues. What can you say about its eigenvectors? *Hint*: there are at least two (linearly independent) eigenvectors that share the same eigenvalue.
10. Could the position matrix X be degenerate? *Hint*: X must have distinct main-diagonal elements, to stand for distinct possible positions.

10.1.2 *Hamiltonian and Energy Levels*

1. Consider the Hamiltonian of the harmonic oscillator (Sections 2.4 and 9.3). Is it Hermitian? *Hint*: X and P are Hermitian and so are X^2, P^2, and indeed the Hamiltonian.
2. Is it a legitimate observable?
3. Can it have an eigenvalue with an imaginary part? *Hint*: no. A Hermitian matrix can have real eigenvalues only.

4. Can it have a negative eigenvalue? *Hint*: no. The number operator must have a nonnegative expectation (Sections 9.2.3 and 9.2.4).
5. What is the minimal eigenvalue of the Hamiltonian?
6. Can it be zero? *Hint*: it must be bigger than the minimal eigenvalue of the number operator, which is zero.
7. What is the physical meaning of this eigenvalue? *Hint*: this is vacuum energy: the minimal energy of the harmonic oscillator.
8. Can the harmonic oscillator have no energy at all? *Hint*: no! Its minimal energy is positive.

10.1.3 *Ground State and Its Conservation*

1. In the Hamiltonian of the harmonic oscillator, look again at the minimal eigenvalue. What is the corresponding eigenvector? *Hint*: the ground state: the state of minimal energy.
2. How does it look like? *Hint*: a (real) Gaussian (Figure 9.1).
3. At the ground state, what is the probability to have a particular amount of energy? *Hint*: at probability 1, it has the minimal energy. It can't have any more. This is deterministic.
4. Can the ground state change dynamically in time? *Hint*: it can only precess, with no physical effect.
5. Why? *Hint*: energy must remain at its minimum.
6. How about any other eigenvector? Can it change dynamically in time? *Hint*: same answer: it can only precess, with no physical effect.
7. Why? *Hint*: energy must remain the same eigenvalue.
8. How much energy may the harmonic oscillator have? *Hint*: the allowed energy levels are the eigenvalues of the Hamiltonian: $\omega\bar{h}(k + 1/2)$ $(k \geq 0)$.
9. Is there a maximal energy? *Hint*: you can raise eigenvalues, time and again, forever. After all, you can never hit the zero vector (Section 9.2.7).

10.1.4 *Coherent State and Its Dynamics*

1. In a coherent state, what is the probability to have a particular amount of energy? *Hint*: this is given in the discrete Poisson distribution (Section 2.5.3 and Figure 9.3).

2. Is this deterministic? *Hint*: no! Each energy level has its own (nonzero) probability.
3. Is this probability constant in time? *Hint*: yes! Each energy level only precesses, with no effect on its probability.
4. Is the coherent state constant in time? *Hint*: no! the (momentum) wave function is a superposition of many energy levels, each precessing at its own frequency. This leads to a dynamic interference: constructive, then destructive, then constructive again, and so on.
5. Is this physical? *Hint*: yes! This affects position–momentum. Over time, there is a real change in the (momentum) wave function (in not only phase but also magnitude), which affects the probability too.
6. Does this affect the probability to have a particular amount of energy? *Hint*: no! The Poisson distribution is still valid. Indeed, each eigenfunction only precesses: changes phase but not magnitude at all.
7. Could this affect any observable that commutes with the Hamiltonian? *Hint*: no! It shares the same eigenvectors, which only precess, with no effect on probability.
8. What is the physical meaning of this? *Hint*: this observable behaves like total energy: it obeys the same conservation law.
9. What about an observable that doesn't commute with the Hamiltonian? Does it have any real dynamics? *Hint*: yes! It is superposed from many energy levels, each precessing at a different frequency. This leads to a dynamic interference: constructive, then destructive, then constructive again, and so on.
10. Give an example. *Hint*: position–momentum.

10.2 Angular Momentum

10.2.1 *Angular Momentum and Its Eigenvalues*

1. Look at an angular-momentum component (like L_z in Section 9.5.1). What is it? *Hint*: an operator.
2. On what functions does it act? *Hint*: it acts on (3-D) wave functions, defined from $\mathbb{R}^3 \to \mathbb{C}$ (Section 9.4.2).
3. Is it Hermitian?
4. Is it a legitimate observable?

5. What can you say about its eigenvalues? *Hint*: they are real.
6. What can you say about its eigenvectors? *Hint*: if they have different eigenvalues, then they are orthogonal.
7. And what if they share the same eigenvalue (but are still linearly independent)? How to orthogonalize them? *Hint*: use a Gram–Schmidt process (Section 10.1.1).
8. Normalize them, so their square has integral 1 (in 3-D).
9. What are they now? *Hint*: orthonormal.
10. Consider some eigenvalue of L_z. Look at its eigenvector. By now, how does it look like? *Hint*: this is an eigenfunction: defined on $\mathbb{R}^3 \to \mathbb{C}$. Besides, it is already normal: its square has integral 1 (in 3-D).
11. Consider some state: a wave function, defined from $\mathbb{R}^3 \to \mathbb{C}$ (x-, y-, and z-position). Assume that it is normal too (its square has integral 1 in 3-D). How likely is the particle to be at (x, y, z)? *Hint*: take this wave function at (x, y, z), take its absolute value, and square it up. This is the probability density (in 3-D).
12. And how likely is the z-angular momentum to be the same as the above eigenvalue? *Hint*: look at these two functions: the eigenfunction (corresponding to this eigenvalue of L_z), and the wave function. By now, they are both normal (in 3-D). Take their inner product (in 3-D), take its absolute value, and square it up. This is the required probability.
13. Does L_z have eigenvalue zero? *Hint*: yes. Indeed, look at the constant eigenfunction (or any other function invariant under interchanging the x- and y-coordinates: $x \leftrightarrow y$).
14. Is L_z singular? *Hint*: yes: it has a nontrivial null space.
15. Next, look at a positive eigenvalue of L_z, and its eigenfunction.
16. How to modify it, and design a new eigenfunction, with the negative eigenvalue? *Hint*: interchange the x- and y-coordinates: $x \leftrightarrow y$. This way, the eigenvalue picks a minus sign.
17. Show that L_z may have discrete eigenvalues of the form
$$0, \pm\bar{h},\ \pm 2\bar{h},\ \pm 3\bar{h}, \ldots$$
(a finite list). *Hint*: see Sections 9.5.4 and 9.5.5.
18. Show that L_z may also have eigenvalues of the form
$$\pm\frac{\bar{h}}{2},\ \pm\frac{3\bar{h}}{2},\ \pm\frac{5\bar{h}}{2},\ \pm\frac{7\bar{h}}{2}, \ldots$$

(a finite list). *Hint*: use symmetry considerations to make sure that, in this (finite) list, the minimal and maximal eigenvalues have the same absolute value.

10.3 Spin Matrices

10.3.1 *Spin-One*

1. Define three new 3×3 matrices:

$$S_x \equiv i\bar{h} \begin{pmatrix} 0 & 0 & 0 \\ 0 & 0 & -1 \\ 0 & 1 & 0 \end{pmatrix}$$

$$S_y \equiv i\bar{h} \begin{pmatrix} 0 & 0 & 1 \\ 0 & 0 & 0 \\ -1 & 0 & 0 \end{pmatrix}$$

$$S_z \equiv i\bar{h} \begin{pmatrix} 0 & -1 & 0 \\ 1 & 0 & 0 \\ 0 & 0 & 0 \end{pmatrix}$$

(where $i \equiv \sqrt{-1}$ is the imaginary number).
2. These are called spin matrices.
3. Show that these definitions are cyclic under the (right-hand) shift:

$$x \to y \to z \to x.$$

4. Are these matrices Hermitian?
5. Are they legitimate observables?
6. What do they observe? *Hint*: they will observe the boson spin: S_x will observe spin-right, S_y spin-in, and S_z spin-up (Figures 10.1 and 10.2).
7. Calculate their commutator. *Hint*:

$$[S_x, S_y] = i\bar{h}S_z,$$

$$[S_y, S_z] = i\bar{h}S_x,$$

$$[S_z, S_x] = i\bar{h}S_y.$$

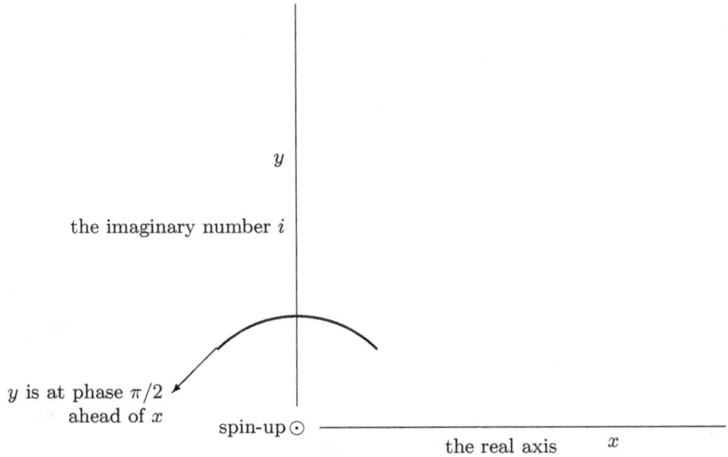

Figure 10.1. The x-y plane: a view from above. The spin vector points in the positive z-direction: from the page toward your eyes (as indicated by "$\odot$" at the origin). Around it, the spin is counterclockwise, as in the right-hand rule. This is spin-up, represented by the eigenvector $(1, i, 0)^t/\sqrt{2}$ (y is at phase $\pi/2$ ahead of x).

8. Are these cyclic too?
9. Are these relations familiar? *Hint*: they mirror the formulas in Section 9.5.2.
10. Conclude that the spin matrices represent angular-momentum components.
11. Focus on S_z. What are its eigenvectors and eigenvalues?
12. Look at $(1, i, 0)^t$. Is it an eigenvector? *Hint*: yes (eigenvalue: $\bar{h}$).
13. What is its geometrical meaning?
14. Interpret it to point in the positive z-direction. *Hint*: its y-component is at phase $90°$ ahead of its x-component (Figure 10.1). Now, follow the right-hand rule: place your right hand with your thumb pointing in the positive x-direction, and your index finger in the positive y-direction. This way, your middle finger will point in the positive z-direction.
15. Next, look at $(1, -i, 0)^t$. Is it an eigenvector too? *Hint*: yes (eigenvalue: $-\bar{h}$).
16. What is its geometrical meaning?
17. Interpret it to point in the negative z-direction (Figure 10.2).
18. Normalize these eigenvectors to have norm 1. *Hint*: divide them by $\sqrt{2}$.

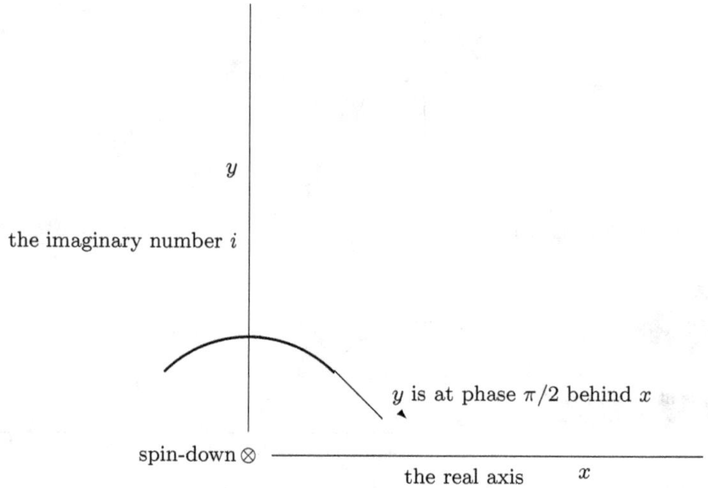

Figure 10.2. In spin-down, on the other hand, the spin vector points down: into the page, away from your eyes (as indicated by "⊗" at the origin). Around it, the spin is counterclockwise again, as in the right-hand rule. This is represented by the other eigenvector: $(1, -i, 0)^t/\sqrt{2}$ (y is at phase $\pi/2$ behind x).

19. Next, look at $(0, 0, 1)^t$.
20. Is it an eigenvector too? *Hint*: yes (eigenvalue: 0).
21. Conclude that it is in the null space.
22. Conclude that S_z is singular.
23. Are these eigenvectors orthogonal? *Hint*: they are eigenvectors of a Hermitian matrix (of distinct eigenvalues).
24. Confirm this explicitly. *Hint*: calculate their inner product. (Don't forget the complex conjugate!)
25. Are they orthonormal? *Hint*: they were already normalized.
26. Is this as in linear algebra? *Hint*: a Hermitian matrix indeed has real eigenvalues and orthogonal eigenvectors.
27. Is this also as expected from Sections 9.5.4 and 9.5.5? *Hint*: yes — an eigenvalue can be raised or lowered by $\bar{h}$.

10.3.2 *The Boson Spin*

1. Consider a new physical system: a boson. This is an elementary particle, with a new property: spin. This is a degenerate kind of angular momentum: it has a direction but no value. As such, it is more mathematical than physical.

2. Interpret S_z as a new observable: spin around the z-axis. This is a degenerate kind of angular momentum: a new random variable, with only three possible results, either counterclockwise (spin-up) or clockwise (spin-down) or none at all.
3. How likely is the boson to spin up? *Hint*: take the state $v \in \mathbb{C}^3$, normalize it to have norm 1, calculate its inner product with $(1, i, 0)^t/\sqrt{2}$, take the absolute value, and square it up. This is the probability to spin up.
4. On the other hand, how likely is the boson to spin down? *Hint*: do the same with $(1, -i, 0)^t/\sqrt{2}$.
5. On the other hand, how likely is the boson to have spin-zero? *Hint*: do the same with $(0, 0, 1)^t$. The result is $|v_3|^2$.
6. Repeat the above exercises for S_x as well. This makes a new observable: spin-right (or left), in the x-direction (around the x-axis).
7. Repeat the above exercises for S_y as well. This makes a new observable: spin-in (or out), in the y-direction (around the y-axis).

10.4 Pauli Matrices

10.4.1 *Spin-One-Half*

1. The above spin is also called spin-one because the maximal eigenvalue is 1 (times $\bar{h}$).
2. Consider now a new physical system: a fermion. (For example, an electron or a proton or a neutron.)
3. This is a simpler particle, with a simpler spin, called spin-one-half because its maximal eigenvalue is 1/2 (times $\bar{h}$).
4. To model this, use a new state, of a lower dimension: $v \in \mathbb{C}^2$.
5. To model this new spin, define three new 2×2 matrices. These are the Pauli matrices:

$$\sigma_z \equiv \begin{pmatrix} 1 & 0 \\ 0 & -1 \end{pmatrix},$$

$$\sigma_x \equiv \begin{pmatrix} 0 & 1 \\ 1 & 0 \end{pmatrix},$$

$$\sigma_y \equiv \begin{pmatrix} 0 & -i \\ i & 0 \end{pmatrix}.$$

6. Are these Hermitian?
7. Are they legitimate observables?
8. What is their trace (sum of main-diagonal elements)? *Hint*: zero.
9. In other words, they are trace-free (or traceless).
10. What is their determinant? *Hint*: -1.
11. What is their square? *Hint*: the 2×2 identity matrix.
12. Show that they are cyclic (in terms of product):

$$\sigma_x \sigma_y = i\sigma_z,$$

$$\sigma_y \sigma_z = i\sigma_x,$$

$$\sigma_z \sigma_x = i\sigma_y.$$

13. Conclude that they have an anti-Hermitian product.
14. Conclude that they anti-commute:

$$\sigma_x \sigma_y + \sigma_y \sigma_x = \sigma_y \sigma_z + \sigma_z \sigma_y = \sigma_z \sigma_x + \sigma_x \sigma_z = (0)$$

(the 2×2 zero matrix).

10.4.2 *New Spin Matrices*

1. Multiply the Pauli matrices by $\bar{h}/2$, and define new spin matrices:

$$\tilde{S}_z \equiv \frac{\bar{h}}{2}\sigma_z,$$

$$\tilde{S}_x \equiv \frac{\bar{h}}{2}\sigma_x,$$

$$\tilde{S}_y \equiv \frac{\bar{h}}{2}\sigma_y.$$

2. These are the new spin matrices (for a fermion, not a boson).
3. Calculate their commutator, and show that it is cyclic as well. *Hint*: recall that the original Pauli matrices anti-commute and are cyclic (in terms of their product). Thus,

$$\left[\tilde{S}_z, \tilde{S}_x\right] = i\bar{h}\tilde{S}_y,$$

$$\left[\tilde{S}_x, \tilde{S}_y\right] = i\bar{h}\tilde{S}_z,$$

$$\left[\tilde{S}_y, \tilde{S}_z\right] = i\bar{h}\tilde{S}_x.$$

4. Conclude that the new spin matrices mirror the angular-momentum components (Section 9.5.2).
5. Look at $\tilde{S}_z$. What are its eigenvectors (and eigenvalues)? *Hint*: the standard unit vectors: $(1, 0)^t$ has eigenvalue $\bar{h}/2$, and $(0, 1)^t$ has eigenvalue $-\bar{h}/2$.
6. Are the eigenvectors orthonormal?
7. Why? *Hint*: this is a Hermitian matrix.
8. Are the eigenvalues real?
9. What is their sum? Is it indeed the same as the trace?
10. What is their product? Is it indeed the same as the determinant?
11. Next, look at $\tilde{S}_x$. What are its eigenvectors and eigenvalues? *Hint*: $(1, 1)^t$ has eigenvalue $\bar{h}/2$, and $(1, -1)^t$ has eigenvalue $-\bar{h}/2$.
12. Are the eigenvectors orthogonal?
13. Why? *Hint*: this is a Hermitian matrix.
14. Normalize them to have norm 1. *Hint*: divide them by $\sqrt{2}$.
15. Are the eigenvalues indeed real?
16. What is their sum? Is it the same as the trace?
17. What is their product? Is it the same as the determinant?
18. Are the eigenvalues as in Sections 9.5.4 and 9.5.5? *Hint*: yes — an eigenvalue can be raised or lowered by $\bar{h}$.
19. Next, look at $\tilde{S}_y$. What are its eigenvectors? *Hint*: $(1, i)^t$ and $(1, -i)^t$.
20. Are they indeed orthogonal to one another? *Hint*: calculate their inner product. (Don't forget the complex conjugate!)
21. Normalize them to have norm 1. *Hint*: divide them by $\sqrt{2}$.
22. What are their eigenvalues? *Hint*: $\pm\bar{h}/2$.
23. What is their sum? Is it indeed the same as the trace?
24. What is their product? Is it indeed the same as the determinant?
25. As in spin-one above, interpret these eigenvectors to model spin. *Hint*: see Figures 10.1 and 10.2.
26. Show that the new spin matrices have the same determinant:
$$\det\left(\tilde{S}_z\right) = \det\left(\tilde{S}_x\right) = \det\left(\tilde{S}_y\right) = -\frac{\bar{h}^2}{4}.$$
27. Show that they also have the same square:
$$\tilde{S}_z^2 = \tilde{S}_x^2 = \tilde{S}_y^2 = \frac{\bar{h}^2}{4} I$$
(where I is the 2×2 identity matrix).

28. Does this agree with the eigenvalues calculated above? *Hint*: take an eigenvector, and apply the matrix to it twice.

10.4.3 *Pauli's Exclusion Principle*

1. Consider now both spin and position at the same time. To tell us about both, how should the state look like? *Hint*: it should contain two wave functions: one for position at spin-up and another for position at spin-down.
2. What is the squared norm of this (big) state? *Hint*: sum both squared norms: that of the former wave function plus that of the latter.
3. Consider now a yet bigger physical system: two fermions (say, two electrons). Could they have exactly the same state (same wave function for position at spin-up and also same wave function for position at spin-down)? *Hint*: no! This is Pauli's exclusion principle.

10.4.4 *Toward Polarization*

1. The Pauli matrices can help model not only spin-one-half but also a new physical property, in a new physical system: a photon. (See Sections 11.2 and 11.4.)
2. The photon is a boson. As such, it has a state in $\mathbb{C}^3$, to help model its spin-one.
3. This state gives rise to a yet simpler state in $\mathbb{C}^2$, to help model yet another interesting physical property: polarization.
4. Indeed, the photon is not only a particle but also light: an electromagnetic wave. As such, it travels in some direction: say, upward (in the positive z-direction). At the same time, it also oscillates (moves, forward and backward) in the horizontal x-y plane.
5. To help model this, the Pauli matrices can also serve as new observables. Better yet, use their spin matrices. For example, $\tilde{S}_z$ tells us how likely the photon is to oscillate in the x- or y-direction. In a (normalized) state $v \equiv (v_1, v_2)^t$, the probability to oscillate in the x-direction is $|v_1|^2$, and in the y-direction is $|v_2|^2$.
6. How is this related to the eigenvectors of $\tilde{S}_z$? *Hint*: the inner product of v with $(1, 0)^t$ is v_1 and with $(0, 1)^t$ is v_2.

7. At the same time, $\tilde{S}_x$ tells us how likely the photon is to oscillate obliquely, at angle 45° in between the x- and y-axes. To calculate the probability for this, take the eigenvector $(1, 1)^t/\sqrt{2}$, calculate its inner product with v, take the absolute value, and square it up.
8. Write the result as a formula in terms of v_1 and v_2.
9. And what is the probability to oscillate in the perpendicular direction? *Hint*: do the same but with the other eigenvector: $(1, -1)^t/\sqrt{2}$.
10. Finally, $\tilde{S}_y$ tells us how likely the photon is to make circles in the x-y plane (Figures 10.1 and 10.2). To calculate the probability to make circles counterclockwise, take the eigenvector $(1, i)^t/\sqrt{2}$, calculate its inner product with v (don't forget the complex conjugate), take its absolute value, and square it up.
11. Write the result as a formula in terms of v_1 and v_2.
12. To calculate the probability to make circles clockwise, on the other hand, do the same with the orthonormal eigenvector: $(1, -i)^t/\sqrt{2}$.

Chapter 11

Spin and Polarization: An Algebraic-Geometrical Model

How was quantum mechanics discovered in the first place? We already saw one major experiment: the double-slit experiment (Figure 4.1, extended theoretically in Figure 8.2). But there was yet another important experiment: the Stern–Gerlach experiment. It looked at an electron, and how it behaves in a magnetic field.

Actually, the electron is a little magnet: a magnetic dipole. In classical physics, this would tend to accept (and adopt) the global magnetic field around it: if it spins parallel to the magnetic field, then it would tend to stay this way. If, on the other hand, it spins anti-parallel, then it would tend to flip. For example, if your magnetic field points up (in the positive z-direction), would a random electron be likely to spin-up too? Not in quantum mechanics (as we'll see later)! Besides, here we're interested in its initial state: before entering the magnetic field. How did it spin originally? Up or down? To find out, throw it into a nonuniform magnetic field.

The results were quite surprising. To help explain them, we'll see an algebraic-geometrical model. In a photon, it will help explain not only spin but also polarization. In an electron, on the other hand, it will help design a (theoretical) quantum computer.

How to see spin? Well, we can't see it directly. Fortunately, spin leads to yet another interesting phenomenon: polarization of light. To see it, use a polarizer. Fortunately, this can be modeled, both geometrically and algebraically.

As in spin, in polarization too, we have three measurables: two linear (horizontal and vertical) and one circular. As in spin, they don't commute. For this reason, we can only measure one of them exactly but not the other two. This is the uncertainty principle.

Thanks to our algebraic tools, we can even model a general polarizer: linear or circular or even both. Likewise, in an electron too, we can even model a general spin: not only up-or-down but also in any other direction in 3-D.

This is just an example of a more general system, with more general rules:

- The dynamics follows from the Schrodinger equation.
- The eigenvectors of the Hamiltonian make energy levels (on which energy is conserved).
- An operator that commutes with the Hamiltonian models a conserved property too.

In our case, spin has magnetic energy (whose Hamiltonian is just a Pauli matrix). This is why only one spin is conserved (the one that is parallel to the magnetic field) but not the other two. This will be useful in quantum computing.

11.1 The Stern–Gerlach Experiment

11.1.1 *Potential in an Electrostatic Field*

Let's start from classical electromagnetics and highlight its shortcoming. For this, consider a static charged ball (at the middle of Figure 11.1). It produces an electrostatic field, whose flux is illustrated as outgoing arrows.

This is the force that the ball would exert on an electron. This comes from the ball only: there is no other source. This is why, farther and farther away from the ball, the arrows diverge and spread out, more and more.

How to quantify this? Consider a big sphere, of radius $r > 0$. What is the total flux through it? Thanks to Green's divergence theorem, this is

$$\iint_{\text{sphere}} \text{field} \cdot \mathbf{n}\, dxdy = \iiint \nabla \cdot \text{field}\, dxdydz = \text{charge of ball}$$

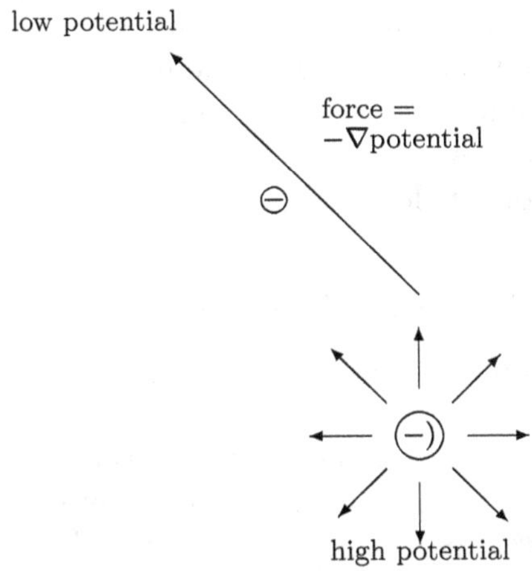

Figure 11.1. The charged ball (at the middle) produces a radial electrostatic field: high potential near the middle, and low potential far away. In this field, an electron travels to the low-potential region, to make its (positive) energy minimal.

(where **n** is the outer normal vector). As a matter of fact, this is true for any closed surface around the ball. Why did we pick a sphere? Because we know its area: $4\pi r^2$. Thus, the flux per unit area (and indeed the field itself) behave like r^{-2}. Fortunately, the field is conservative: has no curl. Thus, it has a potential: its line integral (times -1). This behaves like r^{-1} (plus some constant): maximal near the ball and monotonically decreasing farther and farther away. How would this affect an electron?

11.1.2 *Force: Toward Minimal Energy*

Into this field, throw an electron. What force would it feel? Well,

$$\text{force} = -\nabla \text{potential}.$$

This is radial: outward. Better yet, look at things in terms of energy: an electron has a positive energy that agrees with the field. After all, it could "sit" on an arrow and take a "free ride" to $r = \infty$. It has a positive potential energy: capacity to do work.

Thus, the force could actually be interpreted in terms of energy: the electron tends to travel to a low-potential region:

$$r \to \infty \quad \text{and} \quad r^{-1} \to 0.$$

A proton, on the other hand, behaves the other way around: it has a negative energy that disagrees with the field. This is why it wants to approach the ball and even "kiss" it hello. After all, this is where its negative energy $-r^{-1}$ would get minimal. Let's use the same principle in an electromagnetic field as well.

11.1.3 *An Electron in a Nonuniform Magnetic Field*

Now, instead of an electric field, consider a new magnetic field (Figure 11.2). This time, the arrows point upward: from the north magnetic pole (at the bottom) toward the south magnetic pole (at the top). This is nonuniform: on their way up, the arrows spread out, more and more. As a result, the potential gets lower and lower, all the way up.

Now, into this field, throw two kinds of electrons:

- A spin-up electron aligns with the field. Why? Because it makes a little magnet: a little arrow, whose north head is attracted to the south pole of the global magnetic field. This is why it has a positive magnetic energy. (Indeed, it has a positive eigenvalue with respect

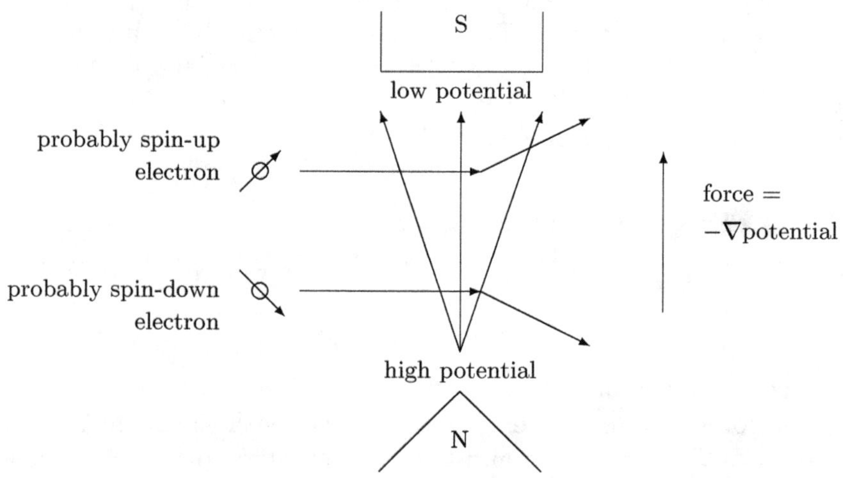

Figure 11.2. An electron in a nonuniform magnetic field. A spin-up electron aligns with the field, and has positive energy. To minimize it, it tends to go up (toward the low-potential region). A spin-down electron, on the other hand, has a negative energy. To minimize it, it tends to go down (toward the high-potential region).

to the Hamiltonian, which is proportional to σ_z.) To minimize it, it tends to go up (to the low-potential region). In other words, it "feels" a magnetic force up.

- A spin-down electron, on the other hand, looks the other way around: south head and north tail. It disagrees with the field and has a negative magnetic energy. (Indeed, it has eigenvalue -1 with respect to σ_z.) To minimize it, it tends to go down, toward the high-potential region (Figure 11.2). There, its energy will get minimal: negative but with a big absolute value.

11.1.4 *The Stern–Gerlach Experiment*

This is the Stern–Gerlach experiment: into a nonuniform magnetic field, fire a stream of electrons (from the left). How should they go? They should go oblique: either strongly up-right or strongly down-right or moderate, in between.

What makes them differ from each other? Well, each electron spins (counterclockwise) around its own spin vector, which points at some angle (from the vertical axis): either 0° (spin-up) or 180° (spin-down) or in between. So, we expect electrons of all kinds: spin-up or spin-down or anything in between. On them, the magnetic force should act strongly (on the two former kinds) or weakly (on the rest). As a result, some should go strongly up, some down, and the rest in between (due to a weaker magnetic force on them).

Surprise surprise: no electron goes in between at all! No electron feels a weak magnetic force at all! All electrons feel a strong force, and go strongly up-right or down-right! Why?

This is quantum mechanics for you: discrete, not continuous. This will be described in our algebraic-geometrical model. It will be good for a photon (light ray) and an electron alike. In a photon, it will help explain spin and polarization at the same time. In an electron, on the other hand, it will be useful in quantum computing, later on.

These are just two examples of a more general system: a general Hamiltonian, with its own energy levels. To realize this, let's go back to our good old (active) style: exercises, followed by hints and solutions. (This is the last time of using this, except for a short chapter at the very end of this book.) Thanks to this, you'll get to rediscover physics for yourself!

11.2 Photon: Spin and Polarization

11.2.1 *Spin Vector and Its Tail*

1. Consider a laser beam, pointing upward (in the positive z-direction). It contains many photons, flying upward (at the speed of light).
2. On its way, the photon not only flies upward but also spins around itself (like the Earth).
3. More precisely, the photon spins around a spin vector, issuing from it (Figure 11.3).
4. If you look from above, what would you see? *Hint*: a point of light, approaching your eyes from below. (Don't try this at home!) Spin is not about movement at all: the laser beam keeps coming from the same source, in the same direction: upward (oscillating on its way, which you can never note).
5. As it propagates upward, how does the electromagnetic wave oscillate? *Hint*: it cycles in the complex plane. This makes it look like a (co)sine in z. But the electromagnetic field is not only a scalar but also a 3-D vector, which also spins horizontally: perpendicular to the direction of propagation. Anyway, neither is

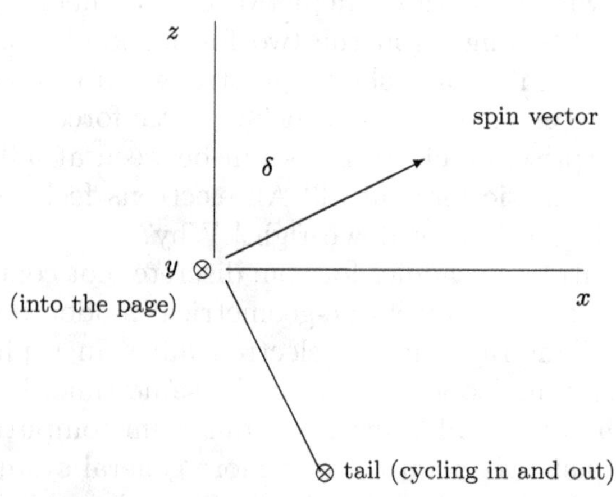

Figure 11.3. The right-hand x-y-z system: z points upward, x rightward, and y deep into the page (away from your eyes, as indicated by "⊗"). The spin vector points up-right (at angle δ from the vertical z-axis). It has a (perpendicular) tail, which cycles around it, in and out: into the page, then out of the page, and so on, round and round (around the spin vector).

about motion but only about the behavior of the electromagnetic field (as a function of z only).

6. In general, each photon may have a different spin vector, pointing in a different direction in 3-D.
7. In a polarized light, on the other hand, all photons share the same spin vector, pointing in parallel, at the same direction in 3-D.
8. The spin vector is constant, and always points in the same direction in 3-D, regardless of time or space.
9. Let's start from classical physics (which is inaccurate, but still interesting).
10. Draw the spin vector: from the photon, toward the up-right. (In our laser beam, assume that this is uniform: for all photons alike.)
11. In classical physics, this is real: the photon really spins around the spin vector.
12. Below the spin vector, draw the (horizontal) x-axis. This way, the spin vector is in the vertical z-x plane. (Later on, we'll extend this yet more.)
13. This way, in classical physics, the spin vector has no y-component at all. (In quantum mechanics, on the other hand, this will mean something else: spin-in is as likely as spin-out: see the following.)
14. Moreover, the spin vector is at angle δ away from the vertical z-axis.
15. For simplicity, normalize it to have norm 1. *Hint*: divide it by its own length.
16. In the vertical z-x plane, the spin vector also has a perpendicular "tail," pointing down-right.
17. For simplicity, normalize it to have norm 1 too. *Hint*: divide it by its own length.
18. But the tail is not static but dynamic: it cycles around the spin vector. Thus, the tail depends on the time $t \geq 0$ (which is proportional to z), but not on x or y.
19. In Figure 11.3, we only see where the tail pointed initially. But this direction is going to change with t.
20. What is the electrical field E? *Hint*: project the tail onto the horizontal x-y plane, and obtain

$$E \equiv E(t) \equiv (E_1(t), E_2(t), 0)^t .$$

21. This way, E has no third component at all: it points horizontal, perpendicular to the laser beam, as required.

11.2.2 *Classical Model*

1. What happens at the beginning? *Hint*: at the beginning, the tail points down-right (and has norm 1, as always). Project it onto the horizontal x-axis. This is E: it points rightward (in the positive x-direction) and has length
$$\|E\| = |\cos(\delta)|.$$
2. What will happen to E later? *Hint*: the tail will cycle around the spin vector, and E will be its (new) horizontal projection.
3. What will happen after a quarter of a period? *Hint*: E will be the same as the tail: both will point in the positive y-direction (into the page, away from your eyes).
4. What will their length be? *Hint*:
$$\|E\| = 1.$$
5. What will happen after half a period? *Hint*: the tail will point up-left, at angle δ from the negative part of the x-axis.
6. What will E be? *Hint*: project the tail onto the negative part of the x-axis.
7. What will its length be? *Hint*:
$$\|E\| = |\cos(\delta)|$$
again.
8. What will happen after three quarters of a period? *Hint*: E will be the same as the tail again: both will point in the negative y-direction (from the page toward your eyes).
9. And what will their length be? *Hint*:
$$\|E\| = 1$$
again, and so on.
10. In the horizontal x-y plane, place a (linear) y-polarizer. (It will let the photon go through only if E aligns with y. Later on, we'll model the polarizer algebraically: Section 11.4.) In yet another

experiment, place an x-polarizer instead. Which one is more likely to let the photon go through? *Hint*: y-polarization (the former) is more likely. More precisely, it is $1/|\cos(\delta)|$ times as likely as x-polarization. Indeed, look at $\|E\|$.

11. To verify this, look at a few simple examples.
12. What happens at $\delta = 0$ (or π)? *Hint*: this is spin-up (or down): the spin vector points up (or down, respectively). This way, E coincides with the tail (all the time): both cycle horizontally (and have length 1, all the time), so x-polarization is as likely as y-polarization.
13. Does this agree with our formula? *Hint*: yes:
$$|\cos(0)| = |\cos(\pi)| = 1.$$
14. On the other hand, what happens at $\delta = \pi/2$? *Hint*: this is spin-right: the spin vector points rightward.
15. In this case, how does the tail behave? *Hint*: it cycles vertically: from the y-axis toward the z-axis: $y \to z$ (as in the right-hand rule).
16. Project it horizontally. How does E look like? *Hint*: E aligns with y and oscillates along the y-axis: in and out.
17. This way, E is one-dimensional:
$$E = (0, E_2, 0)^t .$$
18. What is this? *Hint*: this is linear polarization: only in y and never in x.
19. Does this agree with our formula? *Hint*: yes:
$$|\cos(\pi/2)| = 0.$$
20. As a matter of fact, this is correct even without normalization.

11.2.3 *Circular Polarization*

1. This was spin-right.
2. Later on, we'll also see spin-in: the spin vector points in the positive y-direction (into the page, away from your eyes). In this case, how does the tail behave? *Hint*: it cycles vertically: $z \to x$ (as in the right-hand rule).

3. Better yet, multiply by $i = \sqrt{-1}$, to cycle $x \to -z$. (Spin-right plus spin-in will soon give us a new circular polarization: $x \to y$.)
4. Project it horizontally. How will E look like? *Hint*: E will oscillate along the x-axis: right and left.
5. This way, E will be one-dimensional again:

$$E = (E_1, 0, 0)^t .$$

6. What is this? *Hint*: this is linear polarization: only in x, and never in y.
7. Add both spin-right and spin-in. What is this physically? *Hint*: superposition, or interference.
8. Mathematically, what do you get? *Hint*: E is now two-dimensional:

$$E = (E_1, E_2, 0)^t .$$

9. In the horizontal x-y plane, where does E point? *Hint*: (E_1, E_2).
10. Assume that E_1 is at phase $\pi/2$ ahead of E_2:

$$E_1 = iE_2.$$

11. Thanks to this, how does E look like? In particular, how does its real part look like? *Hint*:

$$(\Re E_1, \Re E_2) = (\Re E_1, \Im E_1).$$

This is how E_1 looks like in the complex plane (Figure 11.4).

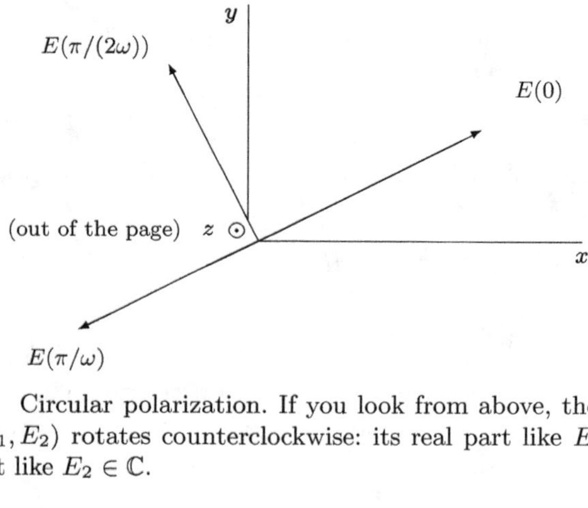

Figure 11.4. Circular polarization. If you look from above, then you'd see E horizontal; (E_1, E_2) rotates counterclockwise: its real part like $E_1 \in \mathbb{C}$, and its imaginary part like $E_2 \in \mathbb{C}$.

12. And how does its imaginary part look like? *Hint*:

$$(\Im E_1, \Im E_2) = (\Re E_2, \Im E_2).$$

This is how E_2 looks like in the complex plane.
13. Give an example. *Hint*: horizontal, round and round (counterclockwise):

$$E(t) = (E_1(0)\exp(i\omega t), E_2(0)\exp(i\omega t), 0)^t .$$

14. In this example, how to guarantee that $E_1 = iE_2$ (all the time)? *Hint*: it is sufficient to require that

$$E_1(0) = iE_2(0).$$

15. In this case, how does E look like? *Hint*:

$$E = E_2(0)\exp(i\omega t)(i, 1, 0)^t.$$

16. What will happen after a quarter of a period, at time $t = \pi/(2\omega)$? *Hint*: E_2 will be as E_1 was initially:

$$E_2\left(\frac{\pi}{2\omega}\right) = iE_2(0) = E_1(0).$$

E_1, on the other hand, will be as E_2 will be later (after half a period):

$$E_1\left(\frac{\pi}{2\omega}\right) = iE_1(0) = i^2E_2(0) = -E_2(0) = E_2\left(\frac{\pi}{\omega}\right)$$

(Figure 11.4).
17. What is this? *Hint*: circular polarization, in its general form (counterclockwise).
18. Why? *Hint*: like $1 - i \in \mathbb{C}$, $(i, 1, 0)^t$ rotates in the horizontal x-y-plane (at frequency ω).
19. Modify this: after a quarter of a period, stretch $\|E\|$. Then, after half a period, shrink $\|E\|$ back again, and so on.
20. What is this? *Hint*: this is an elliptic polarization.

11.2.4 *Superposition: Interference*

1. Next, let E_1 pick a minus sign: $E_1 = -iE_2$. How does the picture look like? *Hint*: the other way around: clockwise rather than counterclockwise. (To see this, you could interchange: $E_1 \leftrightarrow E_2$ and $x \leftrightarrow y$.)
2. Later on, we'll model these pictures by two twin (orthogonal) vectors:
$$\begin{pmatrix} 1 \\ \pm i \end{pmatrix} \in \mathbb{C}^2.$$
3. Moreover, we'll even superpose them (with some coefficients).
4. More concretely, denote these coefficients by $a, b \in \mathbb{C}$.
5. In our superposition, spin-up will have coefficient a, and spin-down coefficient b.
6. More explicitly, how will our superposition look like? *Hint*:
$$a \begin{pmatrix} 1 \\ i \end{pmatrix} + b \begin{pmatrix} 1 \\ -i \end{pmatrix}.$$
7. Assume that a and b are different complex numbers of the same magnitude:
$$a \neq \pm b \quad \text{and} \quad |a| = |b| > 0.$$
8. This way, a and b differ in phase only: a makes angle $\arg(a)$ with the real axis and b angle $\arg(b)$.
9. What does this mean physically? *Hint*: in our superposition, there is a constant phaseshift: one circular polarization is ahead of the other.
10. In Figure 11.5, for example, spin-up is ahead of spin-down.
11. What is the phaseshift between spin-up and spin-down? *Hint*: in the complex plane, the phaseshift is the angle between a and b:
$$0 \neq \arg(b/a) = \arg(b) - \arg(a) \neq \pm\pi.$$
12. And what is the half-phaseshift? *Hint*: this is half of this angle:
$$\arg(\sqrt{b/a}) = \frac{\arg(b) - \arg(a)}{2}.$$
13. In the complex plane, look at two points: 1 and b/a. What do they have in common? *Hint*: both are on the unit circle.

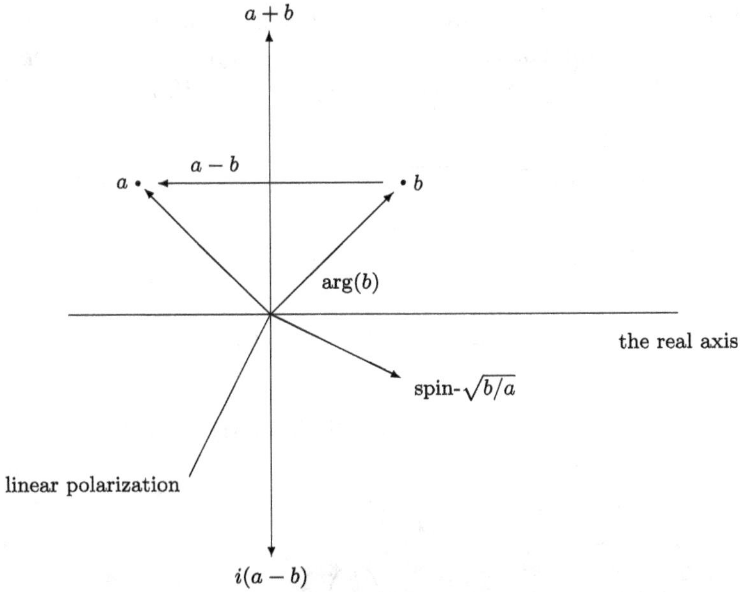

Figure 11.5. In the complex x-y plane, $a + b$ is perpendicular to $a - b$. For example, in our superposition, assume that spin-up (which has coefficient a) is ahead of spin-down (which has coefficient b). This way, $a + b$ and $i(a - b)$ point away from one another and have a negative ratio. Thus, in our superposition, the original circular polarizations interfere and cancel out, yielding spin-$\sqrt{b/a}$, with a perpendicular (linear) polarization.

14. Connect them by a straight line. What is its length? *Hint*: divide this line into two equal parts. For this, issue a new ray from the origin toward $\sqrt{b/a}$. This ray makes angle

$$\arg(\sqrt{b/a}) = \frac{\arg(b) - \arg(a)}{2}$$

with the real axis. Thanks to this,

$$\frac{|a-b|}{|a|} = \frac{|b-a|}{|a|} = \left|\frac{b}{a} - 1\right| = 2\left|\sin\left(\frac{\arg(b) - \arg(a)}{2}\right)\right|.$$

15. Write our superposition yet more explicitly. *Hint*:

$$a\begin{pmatrix}1\\i\end{pmatrix} + b\begin{pmatrix}1\\-i\end{pmatrix} = \begin{pmatrix}a+b\\i(a-b)\end{pmatrix}.$$

16. What is the ratio between the top component (that stands for spin-x or y-polarization or E_2) and the bottom component (that stands for spin-y or x-polarization or E_1)? *Hint*:

$$\begin{aligned}
\frac{a+b}{i(a-b)} &= -i\frac{a+b}{a-b} \\
&= -i\frac{(a+b)\left(\bar{a}-\bar{b}\right)}{|a-b|^2} \\
&= \frac{-i}{|a-b|^2}\left(a\bar{a}-a\bar{b}+b\bar{a}-b\bar{b}\right) \\
&= \frac{-i}{|a-b|^2}\left(|a|^2+2i\Im\left(b\bar{a}\right)-|b|^2\right) \\
&= \frac{2}{|a-b|^2}\Im\left(b\bar{a}\right) \\
&= \frac{2|a|^2}{|a-b|^2}\Im\left(\frac{b}{a}\right) \\
&= \frac{2|a|^2}{|a-b|^2}\sin(\arg(b)-\arg(a)) \\
&= \frac{2\sin\left(\frac{\arg(b)-\arg(a)}{2}\right)\cos\left(\frac{\arg(b)-\arg(a)}{2}\right)}{2\sin^2\left(\frac{\arg(b)-\arg(a)}{2}\right)} \\
&= \operatorname{cotan}\left(\frac{\arg(b)-\arg(a)}{2}\right).
\end{aligned}$$

17. What is so good about this formula? *Hint*: this is a real number (that depends on the phaseshift only).
18. What does this mean physically? *Hint*: E_1 is proportional to E_2, with no phaseshift in between at all.
19. How come? Where is the phaseshift? *Hint*: the phaseshift is only between the original circular polarizations. But they interfere, cancel out, and produce a new linear polarization, at a new (oblique) direction.
20. What is this new direction? *Hint*: in the complex x-y plane, this is spin-$\sqrt{b/a}$ (in between 1 and b/a), or $i\sqrt{b/a}$-polarization. This way, the spin vector is horizontal (with no z-component

at all), and the (linear) polarization is in the perpendicular direction.

21. Assume that arg(a) is bigger than arg(b). In other words, spin-up is ahead of spin-down. (For example, in the complex x-y plane, $a - b$ points leftward, so $i(a - b)$ points downward: Figure 11.5.) This way, in the complex x-y plane, $a + b$ and $i(a - b)$ point away from one another, and have a negative ratio (as indeed in our formula).
22. So far, a and b were complex numbers of the same magnitude. Later on, on the other hand, we'll also pick real a and b (that depend on δ). This will lead to a more complicated (quantum-mechanical) polarization.
23. In general, what is the magnetic field? *Hint*: it is horizontal too, perpendicular to E.
24. And what is its length? *Hint*: proportional to $\|E\|$, all the time.

11.2.5 *Quantum-Mechanical Model*

1. So far, we only looked at classical electromagnetics. Next, let's move on to quantum mechanics (which is more accurate). For this, let's go back to our original spin vector and its tail (Figure 11.3).
2. What is more likely: spin-up (so the tail cycles counterclockwise) or spin-down (so the tail cycles clockwise)? *Hint*: spin-up is more likely. More precisely, it is $\text{cotan}^2(\delta/2)$ times as likely as spin-down.
3. Can you know δ for sure? *Hint*: no! It is nondeterministic.
4. Still, how to estimate δ quite well? *Hint*: in the horizontal x-y plane, place a circular polarizer. Repeat the experiment many times, and gather statistics about many (identical) photons. Count how many went through, and how many didn't. Use this to solve for δ.
5. In the horizontal x-y plane, look at a column vector: the top component stands for spin-x, and the bottom component for spin-y (Section 11.2.4).
6. Model circular polarization by two (orthogonal) column vectors: model counterclockwise cycling by $(1, i)^t$ and clockwise cycling by $(1, -i)^t$. *Hint*: this follows from Section 11.2.3.

7. Are these indeed orthogonal? *Hint*: calculate their inner product in $\mathbb{C}^2$. Don't forget the complex conjugate (say, on top of the latter vector):

$$\begin{aligned}((1,i)^t,(1,-i)^t) &= (1,i)(1,-i)^* = (1,i)(1,i)^t \\ &= 1\cdot 1 + i\cdot i = 1-1 = 0.\end{aligned}$$

8. Normalize them to have norm 1. *Hint*: divide by $\sqrt{2}$.
9. What do you get? *Hint*: these two column vectors make an orthonormal basis, useful to span $\mathbb{C}^2$.
10. Use them to superpose our original state (obtained from our spin vector). *Hint*: sum the following column vectors:

$$\begin{aligned}&\frac{1}{\sqrt{2}}\cos\left(\frac{\delta}{2}\right)\begin{pmatrix}1\\ i\end{pmatrix} + \frac{1}{\sqrt{2}}\sin\left(\frac{\delta}{2}\right)\begin{pmatrix}1\\ -i\end{pmatrix} \\ &= \frac{1}{\sqrt{2}}\begin{pmatrix}\cos\left(\frac{\delta}{2}\right)+\sin\left(\frac{\delta}{2}\right)\\ i\cos\left(\frac{\delta}{2}\right)-i\sin\left(\frac{\delta}{2}\right)\end{pmatrix}.\end{aligned}$$

11. What does this mean physically? *Hint*: interference: constructive in x (the top component), and destructive in y (the bottom component).
12. Look at these two components (the upper and the lower). What is their relative probability? *Hint*: take their absolute value, square it up, and divide:

$$\frac{\left(\cos\left(\frac{\delta}{2}\right)+\sin\left(\frac{\delta}{2}\right)\right)^2}{\left(\cos\left(\frac{\delta}{2}\right)-\sin\left(\frac{\delta}{2}\right)\right)^2} = \frac{1+\sin(\delta)}{1-\sin(\delta)}.$$

13. Thanks to this, we are now ready to estimate linear polarization better. What is more likely: spin-x (so E aligns with y), or spin-y (so E aligns with x)? *Hint*: the former is more likely. More precisely, it is $(1+\sin(\delta))/(1-\sin(\delta))$ times as likely as the latter.
14. Design an experiment to verify this. *Hint*: in the horizontal x-y plane, place a (linear) y-polarizer. In yet another experiment, place an x-polarizer instead. Repeat this many (independent) times, and gather statistics about many (identical) photons.

15. Which one is more likely to let the photon go through? *Hint*: y-polarization (the former) is more likely. More precisely, it should be $(1+\sin(\delta))/(1-\sin(\delta))$ times as likely as x-polarization.
16. In particular, test a few simple examples.
17. For $\delta = 0$ (or π), both x- and y-polarization should be as likely. Why? *Hint*: spin is up (or down), so E cycles in the unit circle in the horizontal x-y plane.
18. Does this agree with our new formula? *Hint*: yes:

$$1 \pm \sin(0) = 1 \pm \sin(\pi) = 1 \pm 0 = 1.$$

19. For $\delta = \pi/2$, on the other hand, things are completely different: only y-polarization is possible. Why? *Hint*: this is spin-right, so E aligns with y, all the time.
20. Does this agree with our new formula? *Hint*: yes:

$$1 \pm \sin(\pi/2) = 1 \pm 1.$$

21. For many more δ's, carry out your experiment, and study the results.
22. Do they agree with our new formula better than with the old (classical) one?
23. Do you accept quantum mechanics now and believe in it?
24. So far, our spin vector was in the vertical z-x plane, with no y-component at all. How to supplement it with a nonzero y-component as well? *Hint*: rotate it horizontally (by angle α), around the vertical z-axis (into the page, away from your eyes). With it, rotate the entire x-y plane (by angle α) as well. For this, apply the new matrix

$$O \equiv \begin{pmatrix} \cos(\alpha) & -\sin(\alpha) \\ \sin(\alpha) & \cos(\alpha) \end{pmatrix}.$$

Now, in the above superposition, substitute $O(1, \pm i)^t$ for $(1, \pm i)^t$, respectively. Superpose these new column vectors (with the same coefficients as before). This will give you the new relative probability: how likely y-polarization is, in comparison with x-polarization (the original coordinates).
25. Write this as a function of α (the angle by which O rotates the entire x-y plane).

26. Can you tell α for sure? *Hint*: no! It is nondeterministic.
27. Still, how to estimate α quite well? *Hint*: repeat the same experiment many (independent) times. Gather statistics about many (identical) photons, and estimate the relative probability between y- and x-polarization (the original coordinates). But you already have this algebraically (as a function of α). So, you can now solve for α.
28. Verify this for an example, in which spin-down is as likely as spin-up (so $\delta = \pi/2$). *Hint*: as before: carry out the experiment many times, and gather statistics about y- and x-polarization (the original coordinates). Their relative probability should be

$$\text{cotan}^2(\alpha) = \frac{\cos^2(\alpha)}{\sin^2(\alpha)}.$$

29. Does this agree with your new formula (once you set $\delta = \pi/2$ in it)?

11.3 On the Electron Spin

11.3.1 *On the Electron Spin*

1. This was a photon. What about an electron? What is its kinetic energy? *Hint*: its kinetic energy is proportional to the inner product of its spin vector with its momentum vector (in 3-D). Indeed, in Dirac's equation, this is the first term.
2. Can these vectors be measured for certain? *Hint*: the momentum vector can because its components commute. But the spin vector can't because its components don't commute. For this reason, only one spin (in one particular direction) can be measured exactly. For example, spin-up and spin-right don't commute, so only one of them can be measured exactly.
3. Still, in the electron too, how to estimate the spin vector quite well? *Hint*: repeat the same experiment many times, and gather statistics about a lot of (identical) electrons, and about their spin vector (one component at a time), and its expectation.
4. In quantum mechanics, time mirrors position, and energy mirrors momentum. Indeed, the momentum wave function is obtained from the Fourier transform (applied to the original wave function).

Likewise, the (discrete) energy wave function is obtained from the Fourier series. Dirac's equation extends this to an electron too (even at a very high speed). For this, it uses special relativity too: momentum–energy mirror each other in yet another way: they form a new relativistic pair, transformed by the Lorentz matrix.
5. This is how special relativity joins forces with quantum mechanics, to form a complete theory for the electron as well.

11.4 Photon and Its Polarizer

11.4.1 *Linear Polarizer*

1. Look again at an electromagnetic wave: a photon, flying upward, in the positive z-direction (Figure 11.3).
2. Look at the horizontal x-y plane from above.
3. While propagating toward your eyes, the wave also oscillates along an oblique line in the x-y plane. We then say that the wave is polarized linearly (or α-polarized: Figure 11.6).
4. Here, the x-oscillation matches the y-oscillation: there is no phaseshift.

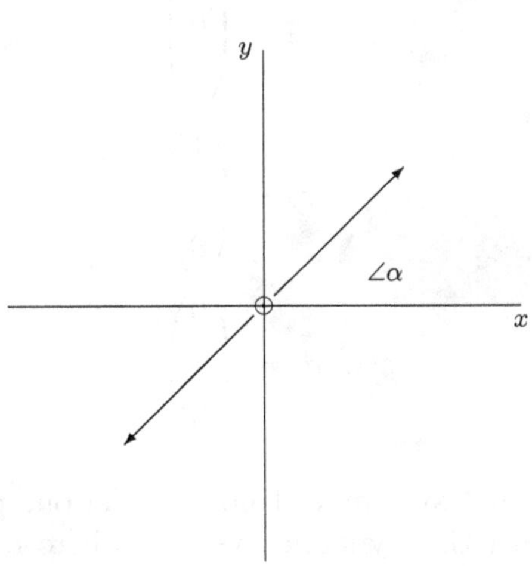

Figure 11.6. A linearly polarized electromagnetic wave: while propagating outward (out of the page, toward your eyes), it also oscillates in the x-y plane, forward and backward, along the oblique line that makes angle α with the positive part of the x-axis.

5. Now, design an α-polarizer: a machine that takes an incoming photon and lets it go through only if it is α-polarized (as in Figure 11.6). It also tells us more: if the incoming photon is η-polarized (at some unknown angle η), then it tells us how likely η was to be exactly α. (α is now the angle of the polarizer, not the photon.) This is done by looking at the light that comes out and measuring how weak it became. In the process, the light also changes forever: it becomes α-polarized (rather than η-).
6. How does the polarizer work mathematically? *Hint*: see the following.
7. Mathematically, the polarizer is an observable: a Hermitian operator (matrix). To design it, use the double angle: 2α (rather than α):

$$\cos(2\alpha)\sigma_z + \sin(2\alpha)\sigma_x = \begin{pmatrix} \cos(2\alpha) & \sin(2\alpha) \\ \sin(2\alpha) & -\cos(2\alpha) \end{pmatrix}.$$

8. Why use the double angle 2α (rather than α)? *Hint*: to add Pauli matrices, we need to mirror them geometrically in 3-D:

$$\sigma_z \sim \begin{pmatrix} 0 \\ 0 \\ 1 \end{pmatrix}$$

$$\sigma_x \sim \begin{pmatrix} 1 \\ 0 \\ 0 \end{pmatrix}$$

$$\sigma_y \sim \begin{pmatrix} 0 \\ 1 \\ 0 \end{pmatrix}.$$

   This (static) 3-D system will help model our polarizer. (Don't confuse it with the dynamic system, where our original wave propagates and oscillates.) Now, in 3-D, the angle must double: opposite spins in $\mathbb{R}^3$ (which differ by 180°) make orthogonal states in $\mathbb{C}^2$ (which differ by 90°).
9. Is this a legitimate observable? *Hint*: is it Hermitian?

10. Does it have an eigenvector of eigenvalue 1? *Hint*: use a column vector that models α-polarization (Figure 11.6):

$$\begin{pmatrix} \cos(2\alpha) & \sin(2\alpha) \\ \sin(2\alpha) & -\cos(2\alpha) \end{pmatrix} \begin{pmatrix} \cos(\alpha) \\ \sin(\alpha) \end{pmatrix}$$
$$= \begin{pmatrix} \cos^2(\alpha) - \sin^2(\alpha) & 2\sin(\alpha)\cos(\alpha) \\ 2\sin(\alpha)\cos(\alpha) & \sin^2(\alpha) - \cos^2(\alpha) \end{pmatrix} \begin{pmatrix} \cos(\alpha) \\ \sin(\alpha) \end{pmatrix}$$
$$= \begin{pmatrix} \cos(\alpha) \\ \sin(\alpha) \end{pmatrix}.$$

11. As a Hermitian matrix, this observable must also have yet another eigenvector, orthogonal to the previous one (Figure 11.7). What is its eigenvalue? *Hint*: -1:

$$\begin{pmatrix} \cos(2\alpha) & \sin(2\alpha) \\ \sin(2\alpha) & -\cos(2\alpha) \end{pmatrix} \begin{pmatrix} -\sin(\alpha) \\ \cos(\alpha) \end{pmatrix}$$
$$= \begin{pmatrix} \cos^2(\alpha) - \sin^2(\alpha) & 2\sin(\alpha)\cos(\alpha) \\ 2\sin(\alpha)\cos(\alpha) & \sin^2(\alpha) - \cos^2(\alpha) \end{pmatrix} \begin{pmatrix} -\sin(\alpha) \\ \cos(\alpha) \end{pmatrix}$$

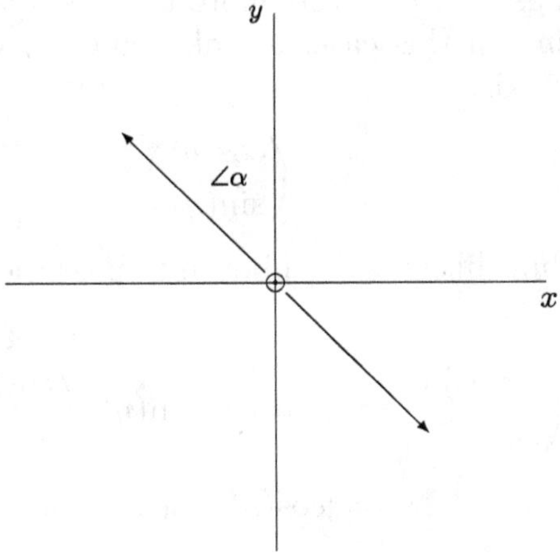

Figure 11.7. Another linearly polarized wave, orthogonal to the previous one: it makes angle α with the positive part of the y-axis (rather than the x-axis).

$$= \begin{pmatrix} \sin(\alpha) \\ -\cos(\alpha) \end{pmatrix}$$

$$= -\begin{pmatrix} -\sin(\alpha) \\ \cos(\alpha) \end{pmatrix}.$$

12. What is the physical meaning of this? *Hint*: the polarizer takes a photon and tells us how it looked like (probably): polarized at angle α (eigenvalue: 1) or $\alpha + \pi/2$ (eigenvalue: -1).
13. Before any experiment was carried out, how likely was the photon to be α-polarized? *Hint*: this depends on its initial (normalized) state: a column vector $v \in \mathbb{C}^2$. Take the inner product of v with the former eigenvector, calculate the absolute value, and square it up:

$$\text{probability to be } \alpha\text{-polarized:} \quad \left| \frac{1}{\|v\|} v^* \begin{pmatrix} \cos(\alpha) \\ \sin(\alpha) \end{pmatrix} \right|^2.$$

14. On the other hand, how likely is the photon to be $(\alpha + \pi/2)$-polarized? *Hint*: in the above inner product, use the latter eigenvector (rather than the former):

$$\text{probability to be } (\alpha + \pi/2)\text{-polarized:} \quad \left| \frac{1}{\|v\|} v^* \begin{pmatrix} -\sin(\alpha) \\ \cos(\alpha) \end{pmatrix} \right|^2.$$

15. For example, assume that the photon was η-polarized. What was its state? *Hint*: in this case, v makes angle η with the positive part of the x-axis:

$$v = \begin{pmatrix} \cos(\eta) \\ \sin(\eta) \end{pmatrix}.$$

16. In this case, how likely is the photon to pass the polarizer? *Hint*: in this case,

$$\begin{aligned} \left| \frac{v^t}{\|v\|} \begin{pmatrix} \cos(\alpha) \\ \sin(\alpha) \end{pmatrix} \right|^2 &= \left| (\cos(\eta), \sin(\eta)) \begin{pmatrix} \cos(\alpha) \\ \sin(\alpha) \end{pmatrix} \right|^2 \\ &= |\cos(\eta)\cos(\alpha) + \sin(\eta)\sin(\alpha)|^2 \\ &= \cos^2(\eta - \alpha) \\ &= \cos^2(\alpha - \eta). \end{aligned}$$

17. Design an experiment to see this probability visually. *Hint*: design a beam of photons, all polarized at the same (unknown) angle: η. Fire the entire beam at our α-polarizer. As before,

probability that η was the same as α:

$$\left| (\cos(\eta), \sin(\eta)) \begin{pmatrix} \cos(\alpha) \\ \sin(\alpha) \end{pmatrix} \right|^2 = \cos^2(\alpha - \eta).$$

Thus, upon passing through, the light intensity should fall by this probability. Only if $\eta = \alpha$ (the probability was 1) should the light pass fully (Figure 11.6). On the other hand, only if $\eta = \alpha \pm \pi/2$ (the probability was 0) should no light pass at all (Figure 11.7). In all other (intermediate) cases, the intensity should fall somewhat and be proportional to the probability.
18. In summary, what is the only relevant angle? *Hint*: $\alpha - \eta$: the angle between the direction of the polarizer and the direction in which the photon was polarized originally.
19. In what case must the photon pass? *Hint*: if $\eta = \alpha$ (probability: 1).
20. In what case can't the photon pass at all? *Hint*: if $\eta = \alpha \pm \pi/2$ (probability: 0).
21. Consider now two polarizers, one by one: an α-polarizer followed by an $(\alpha+\pi/2)$-polarizer. Fire a light beam at them. What comes out? *Hint*: no light at all.
22. Now, in between, place a new $(\alpha+\pi/4)$-polarizer. What will come out now? *Hint*: after passing through the first polarizer, the light gets α-polarized, and its intensity falls somewhat. After passing through the intermediate one, the light gets $(\alpha + \pi/4)$-polarized, and its intensity falls by $\cos^2(\pi/4) = 1/2$. Finally, after passing through the last one, the light gets $(\alpha + \pi/2)$-polarized, and its intensity falls by $\cos^2(\pi/4) = 1/2$ once again.
23. How come? *Hint*: the intermediate polarizer not only measures but also changes the state of the incoming light.
24. Compare this with Bell's experiment: the intermediate measurement changes the state (Figure 8.1).

11.4.2 *General Polarizer: Phaseshift*

1. For simplicity, introduce a new angle:

$$\theta \equiv 2\alpha.$$

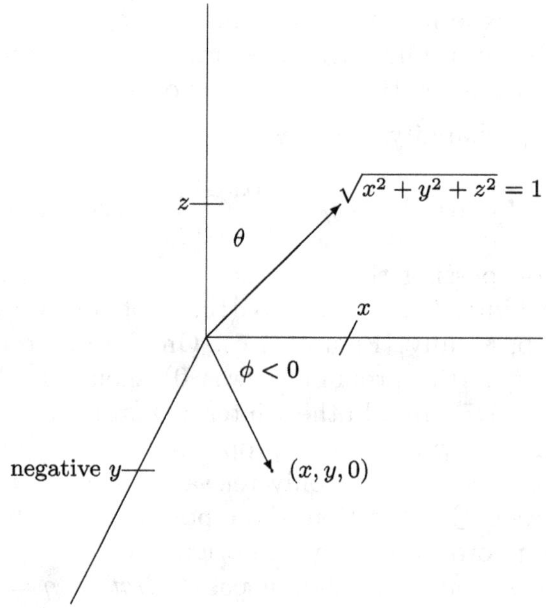

Figure 11.8. A unit vector (x, y, z). It makes angle θ with the vertical z-axis. It also projects a "shadow" $(x, y, 0)$ on the horizontal x-y plane. This shadow makes angle ϕ with the positive part of the x-axis.

2. Modify your polarizer to detect a slightly different polarization: not quite linear, but making an ellipse in the horizontal x-y plane. *Hint*: introduce an angle ϕ, to help draw a new observable:

$$\cos(\theta)\sigma_z + \sin(\theta)\left(\cos(\phi)\sigma_x + \sin(\phi)\sigma_y\right).$$

3. How is ϕ related to δ (Figure 11.3)? *Hint*: $\delta = \phi + \pi/2$ (in the x-z-$-y$ system in Figure 11.8).
4. In our new observable, why use the double angle $\theta = 2\alpha$ (rather than α)? *Hint*: to add Pauli matrices, we must first mirror them in 3-D. More specifically, mirror σ_z, σ_x, and σ_y by the z-, x-, and y-axes, respectively. Now, in 3-D, the angle must double: opposite spins in $\mathbb{R}^3$ (which differ by 180°) make orthogonal states in $\mathbb{C}^2$ (which differ by 90°).
5. Is this a legitimate observable? *Hint*: is it Hermitian?
6. What is the physical meaning of this? *Hint*: in our new polarizer, there is a new phaseshift: the y-polarization (or y-oscillation) is now at phase ϕ ahead of the x-polarization.

7. Show this in the eigenvector. *Hint*: the eigenvector is now

$$\begin{pmatrix} \cos(\alpha) \\ \sin(\alpha)\exp(i\phi) \end{pmatrix}$$

(eigenvalue: 1). Indeed, for $\phi = 0$ (no phaseshift at all), we already saw this before. Thus, for a nonzero ϕ too,

$$\begin{aligned} &(\cos(\theta)\sigma_z + \sin(\theta)(\cos(\phi)\sigma_x + \sin(\phi)\sigma_y)) \begin{pmatrix} \cos(\alpha) \\ \sin(\alpha)\exp(i\phi) \end{pmatrix} \\ &= \begin{pmatrix} \cos(\theta) & \sin(\theta)\exp(-i\phi) \\ \sin(\theta)\exp(i\phi) & -\cos(\theta) \end{pmatrix} \begin{pmatrix} \cos(\alpha) \\ \sin(\alpha)\exp(i\phi) \end{pmatrix} \\ &= \begin{pmatrix} \cos(\alpha) \\ \exp(i\phi)\sin(\alpha) \end{pmatrix}. \end{aligned}$$

8. What is the physical meaning of this? *Hint*: this eigenvector stands for a state of $\cos(\alpha)$ x-polarization and $\sin(\alpha)$ y-polarization (at phase ϕ ahead).
9. While propagating toward your eyes, how does the wave cycle (oscillate)? *Hint*: counterclockwise (along our ellipse), in the horizontal x-y plane.
10. What is the orthogonal eigenvector? *Hint*:

$$\begin{pmatrix} -\sin(\alpha) \\ \cos(\alpha)\exp(i\phi) \end{pmatrix}.$$

11. What is its eigenvalue? *Hint*: -1. Indeed, for $\phi = 0$ (no phaseshift at all), we already saw this before. Thus, for a nonzero ϕ too,

$$\begin{aligned} &(\cos(\theta)\sigma_z + \sin(\theta)(\cos(\phi)\sigma_x + \sin(\phi)\sigma_y)) \begin{pmatrix} -\sin(\alpha) \\ \cos(\alpha)\exp(i\phi) \end{pmatrix} \\ &= \begin{pmatrix} \cos(\theta) & \sin(\theta)\exp(-i\phi) \\ \sin(\theta)\exp(i\phi) & -\cos(\theta) \end{pmatrix} \begin{pmatrix} -\sin(\alpha) \\ \cos(\alpha)\exp(i\phi) \end{pmatrix} \\ &= -\begin{pmatrix} -\sin(\alpha) \\ \exp(i\phi)\cos(\alpha) \end{pmatrix}. \end{aligned}$$

12. What is the physical meaning of this? *Hint*: this eigenvector stands for a state of $-\sin(\alpha)$ x-polarization and $\cos(\alpha)$ y-polarization (at phase ϕ ahead).

13. While propagating toward your eyes, how does the wave cycle (oscillate)? *Hint*: counterclockwise (along another ellipse), in the horizontal x-y plane.
14. How to make it cycle clockwise (rather than counterclockwise)? *Hint*: let ϕ pick a minus sign: $\phi \leftarrow -\phi$.

11.5 Electron in a Magnetic Field

11.5.1 *An Electron and Its Spin*

1. So far, we modeled a light ray (or a photon).
2. Let's use our algebraic-geometrical tools to model an electron too (or any other fermion).
3. To see the geometry, let's start from 3-D.
4. For this, assume that we already have a detector, pointing at a fixed (normalized) direction u:

$$u \equiv \begin{pmatrix} u_1 \\ u_2 \\ u_3 \end{pmatrix} \in \mathbb{R}^3, \quad \|u\| = 1.$$

5. Look at an electron of spin vector w ($\|w\| = 1$ too). In classical physics, the electron "really" spins (counterclockwise) around w. In quantum mechanics, on the other hand, w is nondeterministic and only tells us about probabilities.
6. How likely is the electron to spin-up in the u-direction (or cycle counterclockwise around u)? *Hint*: this depends on the angle θ between u and w (see the following).
7. But θ is often unknown. So, let's leave 3-D for now and go back to $\mathbb{C}^2$, where the original state is defined.
8. In the process, θ gets halved to $\theta/2 = \alpha$.
9. Thus, the probability to spin (counterclockwise) around u is $\cos^2(\alpha)$.
10. Still, this is rather theoretical. After all, α is unknown too. How to calculate this probability in practice? *Hint*: see the following.
11. Look at the observable

$$u_1\sigma_x + u_2\sigma_y + u_3\sigma_z.$$

12. Calculate its eigenvector of eigenvalue 1.

13. Normalize it to have norm 1 in $\mathbb{C}^2$.
14. Take its inner product with the (normalized) state of the electron in $\mathbb{C}^2$.
15. Take the absolute value of this inner product, and square it up.
16. This is the required probability.
17. What is the geometrical meaning of this? Give an example. *Hint*: look at things in $\mathbb{R}^3$ again (see the following).
18. Assume that, in $\mathbb{R}^3$, u and w are perpendicular to each other:

$$\theta = \frac{\pi}{2}.$$

What does this mean physically? *Hint*: the electron has a neutral state: it is equally likely to spin up or down in u (counterclockwise or clockwise around u).
19. Does this agree with our formula? *Hint*: yes:

$$\cos^2(\alpha) = \cos^2\left(\frac{\pi}{4}\right) = \frac{1}{2}.$$

Thus, the electron is indeed equally likely to spin clockwise or counterclockwise around u.

11.5.2 *An Electron in a Uniform Magnetic Field*

1. Next, let's go back to 3-D.
2. In the process, the angle doubles: from $\alpha = \theta/2$ to θ back again.
3. This will give us the full geometrical picture: the electron and its spin in 3-D.
4. Still, this is quite nondeterministic: only one spin can be measured exactly, but not any other spin in any other direction.
5. In 3-D, we can now consider a (uniform) magnetic field, pointing upward (in the positive z-direction):

$$\mathbf{B} \equiv \|\mathbf{B}\| \begin{pmatrix} 0 \\ 0 \\ 1 \end{pmatrix}$$

(Figure 11.9).

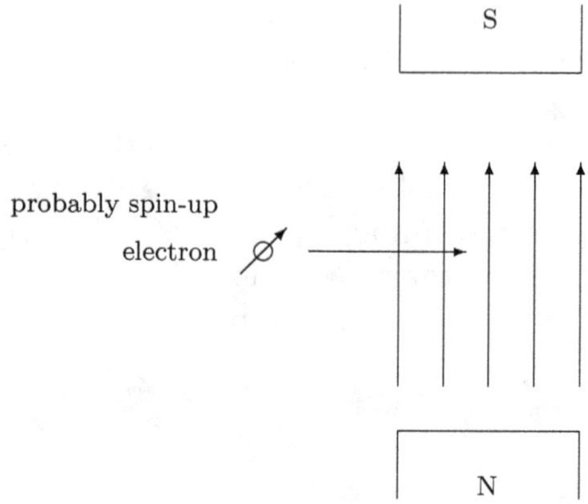

Figure 11.9. The uniform magnetic field $\mathbf{B} \equiv \|\mathbf{B}\|(0,0,1)^t$, pointing upward (in the positive z-direction). The spin vector (of the incoming electron) points at angle θ away.

6. What is the Hamiltonian of this magnetic energy? *Hint*: a 2×2 diagonal matrix, proportional to a spin matrix (and a Pauli matrix):

$$H = \|\mathbf{B}\|\tilde{S}_z = \|\mathbf{B}\|\frac{\bar{h}}{2}\sigma_z = \|\mathbf{B}\|\frac{\bar{h}}{2}\begin{pmatrix} 1 & 0 \\ 0 & -1 \end{pmatrix}.$$

7. Thanks to this (and the Schrodinger equation), the eigenvectors $(1,0)^t$ and $(0,1)^t$ will precess (at opposite frequencies).
8. Look at an electron, entering the field from the left.
9. Let's guess a dynamic (time-dependent) state for our electron spin:

$$v \equiv v(t) \equiv \begin{pmatrix} \cos(\alpha)\exp(-i\phi t) \\ \sin(\alpha)\exp(i\phi t) \end{pmatrix}.$$

(Compare this to the eigenvector of a general polarizer: Section 11.4.2.)
10. Later on, we'll see that this is much more than just a guess: it actually solves the Schrodinger equation.
11. Here, ϕ is the Larmor frequency. It is proportional to $\|\mathbf{B}\|$.
12. How does $v \equiv v(t)$ change and evolve? *Hint*: it precesses at frequency ϕ.
13. Still, for simplicity, we often write just v, not $v(t)$.

14. What is the norm of v? *Hint*: 1, as required.
15. In v, what is the phaseshift between the top and bottom component? *Hint*: $2\phi t$.
16. Does this have any physical effect? *Hint*: not on spin-up (or down). Indeed, the probability for this depends on the top (or bottom) component only, which never changes (in magnitude). To spin in any other direction, however, the probability comes from both. (In spin-right, for example, it comes from their sum.) To this, both contribute and interfere dynamically: constructively, then destructively, then constructively again, and so on.
17. When is the phaseshift exactly π? *Hint*: at time $\pi/(2\phi)$.
18. By this time, what happened to v? *Hint*: by this time, v changed a little: its bottom component picked a minus sign. (Don't worry about an overall phaseshift that has no physical effect at all.)
19. Algebraically, what happened to v? *Hint*: v was multiplied by σ_z:

$$v\left(\frac{\pi}{2\phi}\right) = \sigma_z v(0)$$

(up to an outer phaseshift, with no effect on the probabilities).
20. In our magnetic field, how likely is the electron to spin up (in the positive z-direction)? *Hint*: in v, look at the top component, and square it up:

$$\left|v^t \begin{pmatrix} 1 \\ 0 \end{pmatrix}\right|^2 = |\cos(\alpha)|^2 = \cos^2(\alpha).$$

21. And how likely is it to spin down? *Hint*: do the same with the bottom component:

$$\left|v^t \begin{pmatrix} 0 \\ 1 \end{pmatrix}\right|^2 = \sin^2(\alpha).$$

11.5.3 *Expected Spin-Up*

1. At v, what is the expected spin-up? *Hint*: the relevant observable is a spin matrix, proportional to a Pauli matrix:

$$\tilde{S}_z = \frac{\bar{h}}{2}\sigma_z = \frac{\bar{h}}{2}\begin{pmatrix} 1 & 0 \\ 0 & -1 \end{pmatrix}.$$

This is a diagonal 2×2 matrix. Thus, the expectation (at v) is simply

$$\begin{aligned}
&\left(v, \tilde{S}_z v\right) \\
&= v^* \tilde{S}_z v \\
&= \frac{\bar{h}}{2} \bar{v}^t \sigma_z v \\
&= \frac{\bar{h}}{2} (\cos(\alpha) \exp(i\phi t), \sin(\alpha) \exp(-i\phi t)) \begin{pmatrix} 1 & 0 \\ 0 & -1 \end{pmatrix} \\
&\times \begin{pmatrix} \cos(\alpha) \exp(-i\phi t) \\ \sin(\alpha) \exp(i\phi t) \end{pmatrix} \\
&= \frac{\bar{h}}{2} \left(\cos^2(\alpha) - \sin^2(\alpha)\right) \\
&= \frac{\bar{h}}{2} \cos(\theta).
\end{aligned}$$

2. Does this depend on time? *Hint*: no. It remains the same, all the time.
3. During this calculation, what happened to the angle? *Hint*: it doubled: from $\alpha = \theta/2$ to θ back again, as required in 3-D.

11.5.4 *Expected Spin-Right*

1. Next, what is the expected spin-right (in the positive x-direction)? *Hint*: here, the relevant observable is $\tilde{S}_x = \bar{h}\sigma_x/2$:

$$\begin{aligned}
&\left(v, \tilde{S}_x v\right) \\
&= v^* \tilde{S}_x v \\
&= \frac{\bar{h}}{2} \bar{v}^t \sigma_x v \\
&= \frac{\bar{h}}{2} (\cos(\alpha) \exp(i\phi t), \sin(\alpha) \exp(-i\phi t)) \begin{pmatrix} 0 & 1 \\ 1 & 0 \end{pmatrix} \\
&\times \begin{pmatrix} \cos(\alpha) \exp(-i\phi t) \\ \sin(\alpha) \exp(i\phi t) \end{pmatrix}
\end{aligned}$$

$$
\begin{aligned}
&= \frac{\bar{h}}{2}\left(\cos(\alpha)\sin(\alpha)\exp(2i\phi t) + \sin(\alpha)\cos(\alpha)\exp(-2i\phi t)\right) \\
&= \frac{\bar{h}}{2} 2\sin(\alpha)\cos(\alpha)\,\frac{\exp(2i\phi t) + \exp(-2i\phi t)}{2} \\
&= \frac{\bar{h}}{2}\sin(\theta)\cos(2\phi t).
\end{aligned}
$$

2. During this calculation, what happened to the angles? *Hint*: they doubled: from $\alpha = \theta/2$ to θ, and from ϕ to 2ϕ, as required in 3-D.

11.5.5 *Expected Spin-In*

1. Next, what is the expected spin-in (deep into the page, in the positive y-direction)? *Hint*: here, the relevant observable is $\tilde{S}_y = \bar{h}\sigma_y/2$:

$$
\begin{aligned}
&\left(v, \tilde{S}_y v\right) \\
&= v^* \tilde{S}_y v \\
&= \frac{\bar{h}}{2}\bar{v}^t \sigma_y v \\
&= \frac{\bar{h}}{2}\left(\cos(\alpha)\exp(i\phi t), \sin(\alpha)\exp(-i\phi t)\right)\begin{pmatrix} 0 & -i \\ i & 0 \end{pmatrix} \\
&\quad \times \begin{pmatrix} \cos(\alpha)\exp(-i\phi t) \\ \sin(\alpha)\exp(i\phi t) \end{pmatrix} \\
&= \frac{\bar{h}}{2}\left(\cos(\alpha)\sin(\alpha)(-i)\exp(2i\phi t) + \sin(\alpha)\cos(\alpha)\, i\exp(-2i\phi t)\right) \\
&= \frac{\bar{h}}{2} 2\sin(\alpha)\cos(\alpha)\,\frac{\exp(2i\phi t) - \exp(-2i\phi t)}{2i} \\
&= \frac{\bar{h}}{2}\sin(\theta)\sin(2\phi t).
\end{aligned}
$$

2. During this calculation, what happened to the angles? *Hint*: they doubled: from $\alpha = \theta/2$ to θ, and from ϕ to 2ϕ, as required in 3-D.
3. What is the physical meaning of $2\phi t$? *Hint*: this is the phaseshift (of the bottom component in v, ahead of the top one).

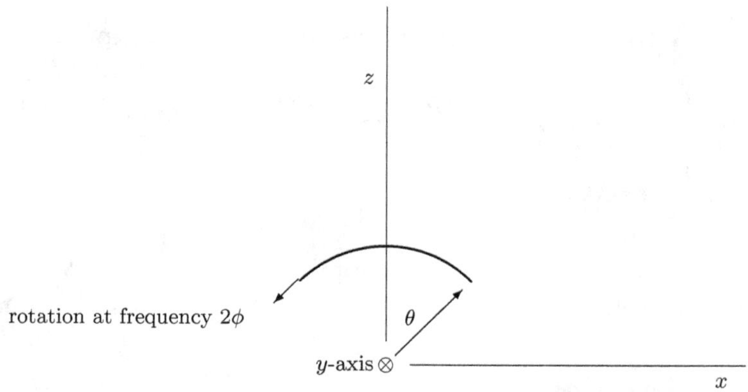

Figure 11.10. Larmor precession: a view from the side. The expected spin vector precesses around the z-axis (at frequency 2ϕ). This is not how the electron moves: it keeps flying rightward but spins up the new spin vector.

11.5.6 *Larmor Precession*

1. By now, we have three expectations. Plug them into a new row vector (Figure 11.10):

$$\frac{\bar{h}}{2}(\sin(\theta)\cos(2\phi t), \sin(\theta)\sin(2\phi t), \cos(\theta)).$$

2. How does this look like? *Hint*: this is a rotation around the z-axis: at angle θ away, and at frequency 2ϕ.
3. This is Larmor's precession.
4. Note: this is not how the electron moves (or its position), but only its spin vector (and where it is expected to point in 3-D).
5. In classical electromagnetics, a spinning charged sphere also behaves in the same way.
6. This is a special case of a more general theorem: the expectations in quantum mechanics obey the same laws as in classical physics.

11.6 Dynamics: Schrodinger's Equation in General

11.6.1 *Hamiltonian and Schrodinger's Equation*

1. In a quantum-mechanical system, consider a state v. Algebraically, what is v? *Hint*: a complex wave function, or a complex vector (of norm 1).

2. Assume now that $v \equiv v(t)$ is time-dependent. How does it evolve in time? *Hint*: it must solve the Schrodinger equation:

$$v'(t) \equiv \frac{dv}{dt}(t) = -\frac{i}{\hbar}Hv(t),$$

where H is the Hamiltonian of the system.
3. How does $v(t)$ look like? *Hint*: in terms of the initial state $v(0)$, it takes the form

$$v(t) = \exp\left(-\frac{i}{\hbar}Ht\right)v(0)$$

(see the following).

11.6.2 *Evolution Operator and Its Diagonal Form*

1. What is the evolution operator that produces this kind of dynamics? *Hint*: $\exp\left(-iHt/\hbar\right)$.
2. This is also called the Liouville operator.
3. How to write it more explicitly? *Hint*: see the following.
4. How to diagonalize it? *Hint*: see the following.
5. Write H in its diagonal form:

$$H = ODO^*,$$

where O is unitary and D is real and diagonal.
6. Look at the elements on the main diagonal of D. What are they? *Hint*: the eigenvalues of H.
7. What are they physically? *Hint*: the energy levels.
8. What are the columns of O? *Hint*: the eigenvectors of H. For instance, let $u \equiv (1, 0, 0, \ldots, 0)^t$ be the first standard unit vector. Then, Ou is the first column in O. This is an eigenvector of eigenvalue $D_{1,1}$:

$$HOu = ODO^*Ou = ODu = D_{1,1}Ou.$$

9. What is this physically? *Hint*: Ou is a state of constant energy level: $D_{1,1}$.
10. Use the above to write the evolution operator more explicitly. *Hint*: thanks to the diagonal form of H, the evolution operator

has its own diagonal form as well:

$$\begin{aligned}\exp\left(-\frac{i}{\bar{h}}Ht\right) &= \sum_{j=0}^{\infty}\frac{1}{j!}\left(-\frac{i}{\bar{h}}Ht\right)^j\\ &= \sum_{j=0}^{\infty}\frac{1}{j!}\left(-\frac{i}{\bar{h}}ODO^*t\right)^j\\ &= \sum_{j=0}^{\infty}\frac{1}{j!}O\left(-\frac{i}{\bar{h}}Dt\right)^j O^*\\ &= O\left(\sum_{j=0}^{\infty}\frac{1}{j!}\left(-\frac{i}{\bar{h}}Dt\right)^j\right)O^*\\ &= O\exp\left(-\frac{i}{\bar{h}}Dt\right)O^*.\end{aligned}$$

11. What did we do here? *Hint*: we pulled both O and O^* out of the exponent function.
12. Conclude that the evolution operator is unitary. *Hint*: since D is real and diagonal, $\exp(-iDt/\bar{h})$ must be unitary: it is diagonal too, with main-diagonal elements of magnitude 1.
13. Is this good? *Hint*: thanks to this, $v(t)$ keeps norm 1, as required.
14. Apply the evolution operator to the initial state $v(0)$. *Hint*: this produces the new state

$$v(t) = \exp\left(-\frac{i}{\bar{h}}Ht\right)v(0) = O\exp\left(-\frac{i}{\bar{h}}Dt\right)O^*v(0).$$

15. On the right-hand side, look at the vector $O^*v(0)$. In it, look at the first component (say). What is this? *Hint*: this is the inner product of $v(0)$ with Ou (the first column of O or the first state of constant energy).
16. How does this individual component evolve in time? *Hint*: it only precesses (clockwise), at frequency $D_{1,1}/\bar{h}$.
17. And how about the other components? How do they evolve?
18. Use this to interpret the above formula better. *Hint*: in the above formula, $v(0)$ was decomposed in terms of columns of O (states

of constant energy). In this decomposition, the coefficients are stored in $O^*v(t)$ $(t \geq 0)$. This way, their sum-of-squares is still 1, as required. As time goes by, each coefficient precesses (at its own frequency), changing phase but not magnitude. This way, the sum-of-squares remains 1, all the time. This is the key to conservation laws.

11.6.3 *Conservation Laws*

1. Consider now a new observable. Could it have any conserved state of its own? How should it look like? *Hint*: its conserved state (if any) must also be a column of O. Indeed, such a column remains the same, all the time (up to an overall phaseshift). A sum of two columns, on the other hand, wouldn't work. After all, they precess at different frequencies and interfere dynamically: constructively, then destructively, then constructively again, and so on. This would make a dynamic state, with a dynamic observation, changing all the time.
2. How should a conserved observable be related to H? *Hint*: they should commute. This way, they share the same eigenvectors (the columns of O) and even the same diagonal form (with the same O but not necessarily the same D).
3. For example, look at the harmonic oscillator (end of Section 8.2.1). Could momentum be conserved? *Hint*: only if there is no potential at all. In this case, there is no force either, so the momentum remains constant. Indeed, the energy is only kinetic, so H is proportional to P^2 (which of course commutes with P). If, on the other hand, H also contains a potential term (like X^2), then it can't commute with P any more, and momentum is not conserved any more.

11.6.4 *Example: Magnetic Energy*

1. For yet another example, consider the physical system in Figure 11.9 again: an electron in a uniform magnetic field. What is the Hamiltonian? *Hint*: since the magnetic field points in the (positive) z-direction, the magnetic energy is proportional to spin-up. More precisely, it is also proportional to the strength

of the magnetic field:

$$H = \|\mathbf{B}\|\tilde{S}_z = \|\mathbf{B}\|\frac{\bar{h}}{2}\sigma_z = \|\mathbf{B}\|\frac{\bar{h}}{2}\begin{pmatrix} 1 & 0 \\ 0 & -1 \end{pmatrix}.$$

2. What are its eigenvectors? *Hint*: the standard unit vectors: $(1,0)^t$ and $(0,1)^t$.
3. What are their eigenvalues? *Hint*: $\pm\|\mathbf{B}\|\bar{h}/2$.
4. What is the physical meaning of this? *Hint*: a spin-up electron agrees with the magnetic field. This is why it has a positive magnetic energy. A spin-down electron, on the other hand, disagrees. This is why it has a negative magnetic energy and tends to flip. (Compare this with Figure 11.2.)
5. In this example, what is O? *Hint*: the 2×2 identity matrix.
6. Thanks to the Schrodinger equation, how should the initial state evolve in time? *Hint*: as discussed in Section 11.5.2,

$$v \equiv v(t) \equiv \begin{pmatrix} \cos(\alpha)\exp(-i\phi t) \\ \sin(\alpha)\exp(i\phi t) \end{pmatrix}.$$

7. In this solution, what is the phaseshift? *Hint*: $2\phi t = \|\mathbf{B}\|t$.
8. At what frequency does the phaseshift precess? *Hint*: $2\phi = \|\mathbf{B}\|$.
9. This is Larmor's frequency.
10. In this system, what spin is conserved? *Hint*: only spin-up: indeed, $\tilde{S}_z$ commutes with the Hamiltonian (which is also proportional to $\tilde{S}_z$). But spin-right and spin-in are not: $\tilde{S}_x$ and $\tilde{S}_y$ don't commute with the Hamiltonian.

Part IV: Algorithms in Cryptography

So far, we focused on quantum mechanics. In the rest of the book, on the other hand, we'll come to practical algorithms. So far, we already saw a few algorithms in machine learning and image processing (Chapters 1 and 5). We also discussed the fundamental unit of data: the bit. For this, we used a fundamental concept from physics: entropy.

How to introduce entropy in both physics and computing at the same time? For this, we used a new logical structure: the binary tree.

Later on, we'll also use a binary tree for yet another purpose: to model a quantum computer. For this, we'll extend the bit into a new unit: the qubit.

The binary tree is also the basis for useful recursive algorithms, such as the fast Fourier transform (FFT). Later on, we'll see FFT in three versions: recursive, nonrecursive, and even on a quantum computer. This will be particularly useful in cryptography.

In this part, we'll see a practical coding-decoding algorithm: the RSA key exchange. This algorithm is widely used to send secret messages over the internet. But there is a catch: to be safe and secret, the algorithm must use a very long prime number, with million digit or so. This is not so easy to find. How to avoid it? use a *probably* prime number. This is good enough, and much easier to find.

To show that the RSA algorithm indeed works, we need a fundamental theorem from number theory: Fermat's little theorem. To prove it, we use the binomial formula (in modular arithmetic).

How to store a long natural number on the computer? Best use its binary representation (as a binary polynomial). This way, it is particularly suitable for modular arithmetic.

How to multiply long natural numbers? Best use FFT. Later on, we'll also implement FFT on a quantum computer. This is quantum FFT, with an interesting application in cryptography.

Chapter 12

Coding–Decoding: The RSA Key Exchange

How to design algorithms in cryptography? For this, we start from the group of integer numbers: $\mathbb{Z}$. In it, we'll use modular arithmetic. This will help design a practical coding–decoding algorithm: the RSA key exchange. To have a safe algorithm, however, we must have a very long prime number, with million digit or so.

Thanks to the binomial formula, we'll prove Fermat's little theorem. This will help prove that the RSA algorithm indeed works. Finally, instead of a prime number, we'll use a *probably* prime number. This is good enough and much easier to find. Later on, we'll also see how to use quantum computing in cryptography.

12.1 Prime Number

12.1.1 *Prime Number*

What is a prime number? This is a natural number that can never be factored as a (nontrivial) product. In other words, it cannot be divided by any smaller number (evenly, with no remainder). Here are a few examples:

$$2, 3, 5, 7, 11, 13, 17, 19, 23, 29, 31, 37, \ldots .$$

Indeed, 11 could be divided by neither 2 nor 3 nor 5 nor 7. Of course, 11 could be divided by 1 and 11 itself, but this doesn't count.

12.1.2 *Prime Numbers and Their Density*

Consider an elementary problem: find all the prime numbers between 2 and N, where N is a given (large) number. How many are there? Quite a bit: as many as $N/\log N$ (times a constant that is independent of N). This is also denoted by

$$O\left(\frac{N}{\log N}\right).$$

What is the density of prime numbers? This is a ratio: the number of prime numbers divided by the total number of numbers:

$$O\left(\frac{1}{\log N}\right).$$

Thus, if you pick a natural number between 2 and N at random, then it has probability $O(1/\log N)$ to be prime. As N grows, this approaches zero but quite slowly. Still, how to find these prime numbers in practice?

12.1.3 *Finding Prime Numbers: a Naive Algorithm*

The naive algorithm uses mathematical induction: assume that we already found all the prime numbers between 2 and $N - 1$. (This is the induction hypothesis.) Now, scan them one by one, and check whether they divide N or not. If any of them does, then N is nonprime. Otherwise, N is prime.

As a matter of fact, there is no need to scan them all. One could stop at the largest prime number p_N for which

$$p_N^2 \le N.$$

After all, if $N = pq$ for some prime numbers p and q, then

$$\min\left(p^2, q^2\right) \le pq = N,$$

so either p or q was already checked, so there is no need to check the other one.

12.1.4 *The Sieve of Eratosthenes*

In the induction step above, how to check whether a prime number p divides N or not? Divide, and ask: is there any remainder? Still, this is too expensive: even if both N and p are in their (efficient) binary representation, this may still cost as many as

$$\log_2 N \log_2 p$$

bit operations. This is too slow.

To avoid this, better use a different approach: look at things the other way around: from the perspective of p, not N. How? Once a prime number p was found, don't wait: use it immediately to find all its multiples, and dismiss them as nonprimes.

This is the sieve of Eratosthenes. It models the following algorithm. Scan the prime numbers one by one, starting from

$$p = 2, 3, 5, \ldots, p_N.$$

For each prime number p, drop all its multiples. After all, they are nonprime. In the end, this leaves only the prime numbers between 2 and N, as required.

12.2 Congruence: An Equivalence Relation

12.2.1 *Division with Remainder*

Look at the group of integer numbers: $\mathbb{Z}$. Thanks to group and number theory, they enjoy all sorts of mathematical equivalence relations.

Let $n > m > 0$ be two natural numbers. What is the ratio between n and m? By "ratio," we don't mean the simple fraction n/m, but only its integer part: the maximal integer number that doesn't exceed it:

$$\left\lfloor \frac{n}{m} \right\rfloor = \max \left\{ j \in \mathbb{Z} \mid j \leq \frac{n}{m} \right\}.$$

This way, we can now divide n by m with remainder (residual). This means representing n in the following form:

$$n = km + l,$$

where

$$k = \left\lfloor \frac{n}{m} \right\rfloor$$

and l is the remainder (or residual) that is too small to be divided by m:

$$0 \leq l < m.$$

This l has a new name: n modulus m. To be recognized as one object, it is often placed in round parentheses:

$$l = (n \bmod m).$$

12.2.2 *Congruence: Same Remainder*

The above equation could also be written as

$$n \equiv l \bmod m.$$

This means that n and l are related: they are *congruent* modulus m. In what sense? In the sense of having the same remainder:

$$n = km + l,$$
$$l = 0 \cdot m + l.$$

This means that both n and l are indistinguishable modulus m. Why? Because both have the same remainder: l:

$$(n \bmod m) = l = (l \bmod m).$$

In other words, their difference $|n - l|$ can be divided by m (evenly, with no remainder at all):

$$m \mid |n - l|.$$

This is indeed a mathematical equivalence relation: reflexive, symmetric, and transitive (check!).

12.3 Greatest Common Divisor

12.3.1 *Common Divisor*

Consider now the special case in which there is no remainder at all:

$$(n \bmod m) = 0.$$

In this case, we say that m divides n:

$$m \mid n.$$

In general, however, m doesn't divide n. Still, even so, m may share a common divisor with n: a third number that divides both. For example, if both are even, then 2 is a common divisor. Still, there could be a few common divisors. What is the maximal one?

12.3.2 *The Euclidean Algorithm*

How to calculate the greatest common divisor of n and m? Well, there are two possibilities: if m divides n, then m itself is the greatest common divisor. If not, then

$$\begin{aligned}
\text{GCD}(n, m) &= \text{GCD}(km + l, m) \\
&= \text{GCD}(l, m) \\
&= \text{GCD}(m, l) \\
&= \text{GCD}(m, n \bmod m).
\end{aligned}$$

So, instead of the original numbers

$$n > m,$$

we can now work with two smaller numbers:

$$m > l = (n \bmod m).$$

This leads to the (recursive) Euclidean algorithm:

$$\text{GCD}(n, m) = \begin{cases} m & \text{if} \quad (n \bmod m) = 0 \\ \text{GCD}(m, n \bmod m) & \text{if} \quad (n \bmod m) > 0. \end{cases}$$

How to prove this? By mathematical induction on $n = 2, 3, 4, \ldots$ (see the following).

12.3.3 The Extended Euclidean Algorithm

Mathematical induction is quite powerful. It can even extend the greatest common divisor and write it as a linear combination:

$$\text{GCD}(n, m) = an + bm,$$

for some integer numbers a and b (positive or negative or even zero). Indeed, if m divides n, then this is easy:

$$\text{GCD}(n, m) = m = 0 \cdot n + 1 \cdot m,$$

as required. If not, then the induction hypothesis tells us that

$$\text{GCD}(n, m) = \text{GCD}(m, l) = \tilde{a} \cdot m + \tilde{b} \cdot l,$$

for some new integer numbers $\tilde{a}$ and $\tilde{b}$ (positive or negative or even zero). Now, in this formula, substitute

$$l = n - km.$$

This gives

$$\begin{aligned} \text{GCD}(n, m) &= \text{GCD}(m, l) \\ &= \tilde{a} \cdot m + \tilde{b} \cdot l \\ &= \tilde{a} \cdot m + \tilde{b}\,(n - km) \\ &= \tilde{b} \cdot n + \left(\tilde{a} - k \cdot \tilde{b}\right) m. \end{aligned}$$

To have the desired linear combination, just define the new coefficients:

$$\begin{aligned} a &= \tilde{b}, \\ b &= \tilde{a} - k \cdot \tilde{b} = \tilde{a} - \left\lfloor \frac{n}{m} \right\rfloor \tilde{b}. \end{aligned}$$

This is the extended Euclidean algorithm.

12.3.4 The Modular Extended Euclidean Algorithm

In practice, however, there is still a problem: the new coefficients a and b could be negative. Worse, they could be too big and even exceed n. How to avoid this? Easy: throughout the above algorithm (including the recursion), introduce just one change: take modulus n. This will force the new coefficients to remain moderate: between 0 and $n-1$:

$$0 \le a, b < n.$$

Likewise, the induction hypothesis should inherit the same change and confine its own coefficients to the same interval: $[0, n-1]$:

$$0 \le \tilde{a}, \tilde{b} < n.$$

To make this change, we must first reformulate the original algorithm as follows:

$$\begin{aligned} a &= \tilde{b}, \\ s &= \left\lfloor \frac{n}{m} \right\rfloor \tilde{b}, \\ b &= \tilde{a} - s. \end{aligned}$$

This changes nothing yet. But now, it is time for a real change. To confine both a and b to $[0, n-1]$, make just a little change: take modulus n:

$$\begin{aligned} a &= \tilde{b}, \\ s &= \left(\left\lfloor \frac{n}{m} \right\rfloor \tilde{b} \right) \bmod n, \\ b &= \begin{cases} \tilde{a} - s & \text{if } \tilde{a} \ge s \\ \tilde{a} + n - s & \text{if } \tilde{a} < s. \end{cases} \end{aligned}$$

Now, in the induction hypothesis, assume also that

$$0 \le \tilde{a}, \tilde{b} < n.$$

Thanks to this, we also have

$$0 \le a, s, b < n,$$

as required. Furthermore, modulus n, nothing has changed: the algorithm still does the same job. Thus, we got a new linear combination modulus n:

$$\text{GCD}(n, m) \equiv (an + bm) \bmod n.$$

Let's go ahead and use this in practice.

12.4 Modular Arithmetic

12.4.1 *Coprime*

Consider a natural number n. What is a coprime to n? This is another number p (prime or not) that shares no common divisor with n:

$$\mathrm{GCD}(n, p) = 1.$$

But n may have a few legitimate coprimes. Which one to pick? Better pick a moderate one. For this purpose, start from an initial guess, say a small prime number:

$$p \leftarrow 5.$$

Does this p divide n? If not, then pick it as our desired coprime. If, on the other hand, this p divides n, then keep looking:

$$p \leftarrow \text{ the next prime number,}$$

and so on. In at most $(\log_2 n)$ guesses, we'll eventually find a new prime number p that doesn't divide n any more and can therefore serve as its legitimate coprime.

12.4.2 *Modular Multiplication*

Let's consider yet another task. Let n and j be two natural numbers. How to calculate the product

$$nj \bmod m?$$

Unfortunately, both n and j could be very long and contain many digits. Their product nj could be even longer and not easy to store. How to avoid this?

Fortunately, both n and j could be written in the form in Section 12.2.2:

$$n = \left\lfloor \frac{n}{m} \right\rfloor m + (n \bmod m),$$

$$j = \left\lfloor \frac{j}{m} \right\rfloor m + (j \bmod m).$$

What do we have on the right-hand side? Well, look at the former term. This is a multiple of m, which is going to drop anyway.

Thus, instead of nj, better calculate

$$(n \bmod m)(j \bmod m) \bmod m.$$

In other words, take the modulus at the beginning, before starting to multiply. This way, you have moderate numbers from the start: never exceeding $m-1$.

12.4.3 *Modular Power*

Next, use this idea time and again, and calculate a power like

$$n^k \bmod m.$$

Here, even for a moderate k, n^k could be too long. Thanks to the above idea, we can now avoid this: before using n^2, take modulus m:

$$\text{power}(n,k,m) = \begin{cases} 1 & \text{if } k=0 \\ n \bmod m & \text{if } k=1 \\ \begin{array}{c} n \cdot \text{power}(n^2 \bmod m, \\ (k-1)/2, m) \bmod m \end{array} & \text{if } k>1 \text{ and } k \text{ is odd} \\ \text{power}(n^2 \bmod m, k/2, m) & \text{if } k>1 \text{ and } k \text{ is even.} \end{cases}$$

Better yet, before even starting, take modulus m:

$$n \leftarrow (n \bmod m).$$

This avoids the large number n^2 in the first place. Instead, the recursion is applied to the moderate number

$$(n \bmod m)^2 \le (m-1)^2.$$

This way, throughout the entire calculation, all intermediate products are kept moderate and never exceed $(m-1)^2$. To prove this, use mathematical induction on k (see exercises in the following).

Still, this seems a bit too pedantic. After all, the modulus operation could be expensive too. Why apply it so often? Better apply it only when absolutely necessary: when an inner (temporary) variable gets too long to store and use.

12.5 Modular Inverse

12.5.1 *Modular Inverse*

Let p and q be be coprime to one another (but possibly nonprime):

$$\text{GCD}(p, q) = 1.$$

Another important task is to find the inverse of q modulus p, denoted by

$$q^{-1} \bmod p.$$

What is this inverse? It is the unique solution x to the equation

$$qx \equiv 1 \bmod p.$$

Is x indeed unique?

12.5.2 *Set and Its Mapping*

Does x exist? Moreover, is it unique? To prove this, consider the set S, containing the integer numbers from 0 to $p - 1$:

$$S = \{0, 1, 2, \ldots, p - 1\}.$$

Define the new mapping

$$M : S \to S$$

by the modular product

$$M(s) = (qs \bmod p), \qquad s \in S.$$

Our aim is to show that M maps some number to 1. This number will then serve as the desired solution x. To show this, let's look at M and study its properties.

12.5.3 *One-to-One Mapping*

Is M one-to-one? To check on this, consider two numbers $a, b \in S$ (say, $a \geq b$). Assume that

$$M(a) = M(b).$$

This means that

$$q(a-b) \equiv 0 \bmod p.$$

In other words, p divides $q(a-b)$:

$$p \mid q(a-b).$$

Since p shares no common divisor with q, it must divide $a-b$:

$$a \equiv b \bmod p.$$

Thus, M is indeed one-to-one.

12.5.4 *Mapping onto the Set*

By now, we saw that M is one-to-one. As such, it preserves the total number of elements in S:

$$|M(S)| = |S| = p.$$

Thus, M covers S: it is a one-to-one mapping from S onto S. As such, M is reversible.

12.5.5 *The Inverse Mapping*

In other words, M has an inverse mapping:

$$M^{-1} : S \to S.$$

In particular, we can now define

$$x = M^{-1}(1).$$

This is what we wanted to prove: the inverse of q modulus p exists uniquely:

$$\left(q^{-1} \bmod p\right) = M^{-1}(1).$$

12.5.6 *Using the Extended Euclidean Algorithm*

How to solve for x in practice? For this purpose, assume that

$$q < p.$$

(Otherwise, just substitute

$$q \leftarrow q - p$$

time and again, until q gets as small as p. After all, this makes no difference to x.) Now, use the modular extended Euclidean algorithm (Section 12.3.4). This way, we can now write

$$1 = \text{GCD}(p, q) \equiv (ap + bq) \equiv bq \bmod p,$$

for some integer numbers a and b that lie in between 0 and $p - 1$. Obviously, the desired solution is just

$$x = b.$$

In summary, the inverse of q modulus p is now available explicitly:

$$\left(q^{-1} \bmod p\right) = b.$$

This will be useful in the following.

12.6 The RSA Algorithm: Coding

12.6.1 *The Coding Part*

We are now ready for the RSA algorithm: coding-decoding [18]. Assume that the original message was already translated into a long natural number (containing thousands of digits, or even more). Before it is sent over the internet, it must be coded safely. How to do this? Let

$$p < q$$

be two long prime numbers that are kept secret. Their product

$$n = pq,$$

on the other hand, is not secret at all: it is available to the public. Define also

$$\phi = (p - 1)(q - 1).$$

Although n is public, ϕ is not. As a matter of fact, since both p and q are secret, $p - 1$ and $q - 1$ are secret as well, so their product can't be calculated by anyone.

Furthermore, let $e \geq 5$ be a moderate coprime of ϕ. Unlike ϕ, e is not secret: it is placed in the public domain. There is no risk: e is not enough to uncover ϕ.

Now, assume that our message m satisfies

$$m < p.$$

How to send m safely to the decoder? For this purpose, m itself must never be sent: someone could intercept and read it! To avoid this, only a coded version should be sent.

Fortunately, like everyone, the sender has access to both n and e and can use them to prepare the coded message c:

$$c = (m^e \bmod n).$$

This is the coding part. You can now send c to the decoder. Don't worry: nobody can decode it, but only the legitimate decoder, who knows ϕ. For this purpose, he/she will calculate the modular inverse

$$d = \left(e^{-1} \bmod \phi\right).$$

Don't worry: only he/she can do this. After all, nobody else knows ϕ. Later on, d will be used to decode (uncover the original message m). To see this, we'll need a fundamental theorem in number theory: Fermat's little theorem.

12.7 Fermat's Little Theorem

12.7.1 *Binomial Coefficients*

To help prove Fermat's little theorem, we need a simple lemma. Let p be a prime number. For any natural number $0 < i < p$, consider the binomial coefficient:

$$\binom{p}{i} = \frac{p!}{i!(p-i)!}.$$

In this expression, where could p be found? Only in the numerator, not in the denominator! In other words,

$$p \mid \binom{p}{i}.$$

This will be useful in the following.

12.7.2 *Fermat's Little Theorem*

We are now ready to introduce Fermat's little theorem. Let p be a prime number. For any natural number a, the theorem tells us an interesting fact: the pth power has no effect:

$$a^p \equiv a \bmod p.$$

This was proved by Euler.

12.7.3 *Euler's Proof*

The proof is by mathematical induction on a. Indeed, for $a = 1$, we clearly have

$$a^p = 1^p = 1.$$

We are now ready for the induction step. For some $a \geq 1$, assume that

$$a^p \equiv a \bmod p.$$

This is the induction hypothesis. Let's use it in Newton's binomial formula. Indeed, thanks to the above lemma, most binomial coefficients drop (modulus p):

$$(a+1)^p = \sum_{i=0}^{p} \binom{p}{i} a^i \equiv a^p + 1 \equiv a + 1 \bmod p,$$

as asserted. This completes the induction step. This is Euler's proof.

12.7.4 *Corollary*

So far, we proved that

$$p \mid a^p - a = (a^{p-1} - 1)a.$$

On the right-hand side, we have a product of two factors: $a^{p-1} - 1$ times a. As prime, p must divide either a or $a^{p-1} - 1$.

Does p divide a? Assume not. In other words, assume that a is not a multiple of p:

$$p \not\mid a.$$

In other words, assume that a is coprime to p:

$$\mathrm{GCD}(a,p)=1.$$

In this case, p must divide the other factor:

$$p \mid a^{p-1}-1.$$

In other words,

$$a^{p-1}\equiv 1 \bmod p.$$

In particular, this holds whenever $a<p$. After all, in this case, a is indeed coprime to p, as required: This will be useful in the following.

12.7.5 *The Order of a*

This leads to yet another corollary. As before, assume that a is not a multiple of p:

$$\mathrm{GCD}(a,p)=1.$$

Let q be the minimal natural number for which

$$a^q\equiv 1 \bmod p.$$

(q is called the order of a and is denoted by $q=o_p(a)$.) Then, q divides $p-1$:

$$q \mid p-1.$$

Indeed, from the definition of q as minimal,

$$q\le p-1.$$

Let's show that q indeed divides $p-1$. To see this, divide $p-1$ by q (with remainder):

$$p-1=lq+r,$$

where l is some natural number and r is the remainder:

$$0\le r<q.$$

This way,

$$a^r = a^r 1^l \equiv a^r a^{ql} = a^{lq+r} = a^{p-1} \equiv 1 \bmod p.$$

But q was defined as minimal. So, r must vanish:

$$r = 0.$$

This proves that

$$q \mid p - 1,$$

as asserted.

12.8 The RSA Algorithm: Decoding

12.8.1 *The Decoding Part*

We are now ready to introduce the decoding part. For this purpose, we use the same notations as in Section 12.6.1: m is the original message, p and q are (private) prime numbers, etc.

Since $m < p < q$, m is coprime to both. Likewise, every power of m is also coprime to both p and q, and can therefore play the role of a in Section 12.7.4. Thus,

$$m^\phi = \left(m^{p-1}\right)^{q-1} = 1 + Jq,$$

$$m^\phi = \left(m^{q-1}\right)^{p-1} = 1 + Lp$$

for some nonnegative integers J and L, satisfying

$$Jq = Lp.$$

Thus,

$$p \mid Lp = Jq.$$

Since both p and q are primes, p must divide J as well:

$$J = J'p$$

(for some new nonnegative integer J'). Thus,

$$Jq = \left(J'p\right) q = J'(pq) = J'n.$$

This means that

$$m^\phi \equiv 1 \bmod n.$$

By using Fermat's little theorem twice, we managed to get rid of both p and q and obtain a new formula modulus n. We are now ready to decode the coded message c (Section 12.6.1). Don't worry: only the decoder knows d and can use it in a new modular power:

$$c^d \equiv (m^e)^d = m^{ed} = m^{1+K\phi} \equiv m \bmod n$$

(for some new nonnegative integer K). This is indeed decoding: by calculating the modular power $c^d \bmod n$, the decoder uncovers the original message m secretly, as required.

Is this a good code? This depends on yet another question: are p and q safe? Well, to discover them, a hacker has a rather difficult task: to factorize n as $n = pq$. This is not easy: if p and q are picked big enough, then n would be even bigger and too hard to factorize in an acceptable time. No hacker of a right mind would even attempt to do this: they'd probably find something else to do and leave us in peace.

12.9 Probably Prime Number

12.9.1 *Probably Prime Number*

To have a safe coding–decoding algorithm, we need two (very long) prime numbers: p and q. How to find such? This is too difficult. Instead, let them be only *probably* prime. This is still good enough. After all, no hacker of a right mind would waste his/her time on a code that is probably unbreakable. And what about those of a twisted mind? Let's leave them alone.

To find a probably prime number, we must be able to test: given a natural number, is it prime or not?

12.9.2 *The Naive Test*

Given a natural number n, is it prime or not? Consider a naive test: pick some natural number a in between

$$1 < a < n.$$

Now, if

$$a^{n-1} \equiv 1 \bmod n,$$

then the test is passed. Otherwise, the test fails.

How reliable is this test? Well, thanks to Fermat's little theorem, if n happens to be prime, then the test must be passed:

$$n \text{ is prime} \quad \Rightarrow \quad \text{the test is passed.}$$

So, in this direction, the test never lies: if n is prime, then we are going to know this. We'll never miss any prime number.

Still, what about the other way around? Is it also true that

$$\text{the test is passed} \quad \Rightarrow \quad n \text{ is prime?}$$

Unfortunately not. There are nonprime numbers n that do pass the test for too many a's. Such a's are called liars.

So, there are nonprime numbers with too many liars. This makes the test unreliable and unsafe. It is too easy to pass and could mislead us to believe that a nonprime number is prime.

12.9.3 *The Stronger Rabin–Miller Test*

Fortunately, the Rabin–Miller test is safer [16, 17]. This test applies to any odd number $n > 3$. This way, $n - 1$ is even. Filter out powers of 2 from it. In other words, factorize $n - 1$ in the form:

$$n - 1 = 2^r s$$

(where $r > 0$ is maximal, so s is odd). Now, pick some integer number a in between

$$1 < a < n - 1.$$

Next, look at the following factorization:

$$\begin{aligned}
a^{n-1} - 1 \\
&= a^{2^r s} - 1 \\
&= \left(a^{2^{r-1}s} - 1\right)\left(a^{2^{r-1}s} + 1\right) \\
&= \left(a^{2^{r-2}s} - 1\right)\left(a^{2^{r-2}s} + 1\right)\left(a^{2^{r-1}s} + 1\right) \\
&= \left(a^{2^{r-3}s} - 1\right)\left(a^{2^{r-3}s} + 1\right)\left(a^{2^{r-2}s} + 1\right)\left(a^{2^{r-1}s} + 1\right) \\
&= \cdots = (a^s - 1)(a^s + 1)\left(a^{2s} + 1\right)\left(a^{2^2 s} + 1\right)\cdots\left(a^{2^{r-1}s} + 1\right).
\end{aligned}$$

If n divides any of these factors, then the test is passed. In other words, the test is passed if either

$$a^s \equiv 1 \bmod n$$

or

$$a^{2^j s} \equiv (n-1) \bmod n$$

(for any $0 \le j < r$). Otherwise, the test fails.

How reliable is this test? Well, if n happens to be prime, then Fermat's little theorem guarantees that n divides the left-hand side in the above factorization. Therefore, n must also divide at least one factor on the right-hand side. Thus, the test is passed:

$$n \text{ is prime} \quad \Rightarrow \quad \text{the test is passed.}$$

This is good: we are never going to miss any prime number. If n is prime, then we are going to know this.

Still, what about the other way around? Assume now that we don't know whether n is prime or not. Instead, we only know that the test is passed. This means that n divides at least one factor on the right-hand side. Therefore, n also divides the left-hand side. Thus, this test is stronger (more difficult to pass) than the naive test:

$$\text{the test is passed} \quad \Rightarrow \quad \text{the naive test is passed.}$$

Still, is it also true that

$$\text{the test is passed} \quad \Rightarrow \quad n \text{ is prime?}$$

Not quite. Still, the situation is now better than before. Why? Because the Rabin–Miller test is often *strictly* stronger.

The naive test is unreliable: there are nonprime n's that do divide $a^{n-1} - 1$ for too many liar a's. Fortunately, such an n may *not* pass the (stronger) Rabin–Miller test. In fact, for most a's, such an n may fail to divide *any* of the factors on the right-hand side. For example, such an n could be a product of two factors from the above right-hand side (or their subfactors). In such a case, the Rabin–Miller test is *strictly* stronger and tells us the truth: n is nonprime! In this sense, it is better and safer.

Still, the Rabin–Miller test is not absolute: there may be a nonprime n that does pass it for a few liar a's but not too many: at most 25%. Thus, if a is picked at random from the interval $[2, n-2]$, then it is not very likely to be a liar: the probability for this is as low as $1/4$.

To improve on this, repeat the test k independent times (possibly in parallel). Each time, pick a new a at random. If all tests are passed, then declare n as probably prime. After all, a nonprime number is highly unlikely to pass k independent tests. In fact, the probability to pick k independent liars is as low as

$$\left(\frac{1}{4}\right)^k = 4^{-k}.$$

For $k = 10$, for instance, this is as low as

$$4^{-10} = 2^{-20} = 1024^{-2} < 10^{-6}.$$

12.9.4 *How to Break the Code?*

Suppose that you want to use the RSA code. For this, you already picked a probably prime number p. Now, you also need another probably prime number $q > p$. How to pick it?

Better be careful: since the product $n = pq$ is nonprime, it should better *not* pass the naive test for too many liar a's. Otherwise, any hacker could easily break it!

To do this, the hacker could pick such an a and use it to test n with the Rabin–Miller test. After all, although a is a liar for the naive test, it is probably *not* a liar for the Rabin–Miller test: say, p divides just one factor in the above right-hand side, and q another. This is not good: the hacker could just take this factor and calculate its GCD with n. This would uncover p and break our RSA code!

How to avoid this? Better pick q such that $n = pq$ has only a few liars, even with respect to the naive test. This way, no hacker of a right mind would even attempt to break your code. And what about those of a wrong mind? Let them work forever, searching for a special a that is a liar only for the naive test but not for the Rabin–Miller test.

12.10 Exercises

12.10.1 *Modular Power*

1. Show that the relation

$$l \equiv n \bmod m$$

(Sections 12.2.1 and 12.2.2) is indeed a mathematical equivalence relation between l and n. *Hint*: show that it is reflexive, symmetric, and transitive.
2. Prove that the sieve of Eratosthenes indeed works and finds all the prime numbers between 2 and N.
3. How to calculate $n^k \bmod m$ efficiently? *Hint*: see the following.
4. The naive method uses two stages: first, calculate n^k. Then, take the modulus with m. What's wrong with this?
5. In this method, what kind of numbers are used? How large could they be? Could they be too long to store on the computer? Could they be too long to work with?
6. How to avoid this?
7. Look at the formula at the end of Section 12.4.3. Does this avoid the above problem? How?
8. Does this use short numbers only? How short?
9. Does this still give the correct answer? *Hint*: use mathematical induction on $k = 1, 2, 3, \ldots$.
10. Before using this formula, substitute

$$n \leftarrow (n \bmod m).$$

This way, throughout the entire calculation, are there short numbers only? How short?
11. More specifically, throughout the entire calculation, is there any (intermediate) product as large as m^2? Why?
12. Prove that there is none. *Hint*: use mathematical induction on $k = 1, 2, 3, \ldots$. Note that the inner recursion is applied to

$$(n \bmod m)^2 \bmod m < m$$

and can therefore benefit from the induction hypothesis.
13. Still, is this worthwhile? After all, the modulus operation may be quite expensive. How to avoid using it too often? *Hint*: use it only for (intermediate) numbers that are too long to work with.

12.10.2 *Extended GCD*

1. Prove that the Euclidean algorithm (Section 12.3.2) indeed works, and finds the greatest common divisor. *Hint*: use mathematical induction on $n = 2, 3, 4, \ldots$.
2. Prove that the extended Euclidean algorithm (Section 12.3.3) indeed works, and helps write the greatest common divisor as a linear combination. *Hint*: use mathematical induction on $n = 2, 3, 4, \ldots$.
3. In this linear combination, could the coefficients be negative? How to avoid this? *Hint*: see Section 12.3.4.
4. Look at the algorithm in Section 12.4.1. Does it work? Why?
5. Prove that it indeed finds a small coprime, as required.

12.10.3 *Probably Prime Numbers*

1. Consider a prime number n. Must it pass the naive test in Section 12.9.2? *Hint*: this follows from Fermat's little theorem.
2. Assume that n is also odd, and greater than 3. Must it pass the Rabin–Miller test? *Hint*: this follows from Fermat's little theorem and the factorization in Section 12.9.3.
3. And what about the other way around? Assume that we don't know if n is prime or not. Instead, we only know that it passes the naive test. Is it prime? *Hint*: not necessarily: a could be a liar.
4. Assume that this n is also odd and greater than 3. Must it pass the Rabin–Miller test as well? *Hint*: not necessarily: it could be a product of two factors (or their subfactors) from the factorization in Section 12.9.3.
5. And what about the other way around? If n passes the Rabin–Miller test, must it pass the naive test as well? *Hint*: this follows from our factorization.
6. Must it be prime? *Hint*: not necessarily: it could still be non-prime.
7. Is it more likely to be prime?
8. How likely is it to be prime? *Hint*: at probability 3/4 or even more. After all, a was picked at random from the interval $[2, n - 2]$. In this interval, at most 25% are liars for the Rabin–Miller test, and the rest are not.

9. How to increase this probability even more? *Hint*: repeat the test ten times. Each time, pick a new a at random. If all tests are passed, then declare n as probably prime. After all, it is highly unlikely that n was nonprime and that all these (independent) a's were liars.
10. Would any hacker even attempt to break an RSA code that uses probably prime numbers? Why? *Hint*: the code is probably unbreakable, so no hacker would waste time on it.
11. Implement the RSA algorithm on your computer, with probably prime numbers. *Hint*: the solution can be found in Chapter 5 in Ref. [24].

Chapter 13

Fast Fourier Transform and Its Application

In Chapter 6, we already designed a binary tree and used it to model a computer code. Here, we use the same logic to design a practical algorithm: fast Fourier transform (FFT). How is this relevant to cryptography?

In cryptography, we often use very long natural numbers, with thousands of digits, or even millions. How to store them on the computer? Best use their binary representation (as a binary polynomial). After all, in this form, they are particularly easy to multiply. How? By using FFT.

In quantum mechanics, we often use the Fourier transform, in its continuous version: transform a continuous function into a new continuous function. In image processing, on the other hand, we work with digital images: not continuous but discrete. To them, we apply the Fourier transform in its discrete version. This is useful in compression, pattern recognition, and more (Chapter 5). How to do this efficiently? Best use FFT (introduced by Cooley and Tukey).

To introduce FFT in a simple way, best use a new algebraic tool: a polynomial, sampled evenly on the unit circle (in the complex plane). This helps estimate the complexity (total cost) and see how low it is.

Thanks to FFT, we can now calculate convolution fast. Thanks to this, we can now multiply long natural numbers, in four stages:

- Apply FFT.
- Multiply digit by digit.

- Apply inverse FFT, which is nothing but FFT itself.
- Finally, push digits ahead, if necessary.

This is easy enough to implement on the digital computer. Later on, we'll also implement it on a quantum computer: quantum FFT.

13.1 The Discrete Fourier Transform

13.1.1 *Vector and Its Polynomial*

Consider the n-dimensional complex vector

$$f \equiv (f_0, f_1, f_2, \ldots, f_{n-1})^t \in \mathbb{C}^n,$$

with n complex components. Let's use these components as coefficients in a new polynomial:

$$f(z) \equiv f_0 + f_1 z + f_2 z^2 + \cdots + f_{n-1} z^{n-1} = \sum_{i=0}^{n-1} f_i z^i,$$

where z is the complex variable.

13.1.2 *Root of Unity*

How to transform f? Sample $f()$ at the roots of unity (Figure 13.1). These are n distinct points, distributed evenly on the unit circle, in the complex plane. The rightmost point is the complex number 1. The next point above it is

$$w \equiv \exp\left(\frac{2\pi\sqrt{-1}}{n}\right) = \cos\left(\frac{2\pi}{n}\right) + \sqrt{-1}\sin\left(\frac{2\pi}{n}\right).$$

Why is this a root of unity? To see this, raise it to power n:

$$w^n = 1.$$

What is so good about w? Well, look at its powers:

$$\left\{w, w^2, w^3, \ldots, w^n = 1\right\}.$$

These are roots of unity in their own right. Indeed, for $j = 1, 2, 3, \ldots$, look at w^j. Raise it to power n:

$$\left(w^j\right)^n = w^{jn} = w^{nj} = (w^n)^j = 1^j = 1,$$

as required.

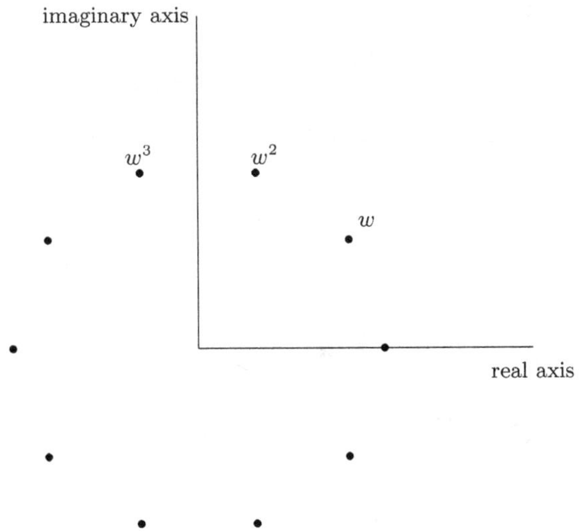

Figure 13.1. In the complex plane $\mathbb{C}$, the roots of unity are $w, w^2, w^3, \ldots,$ $w^n = 1$.

13.1.3 *The Discrete Fourier Transform*

What does the discrete Fourier transform do? It transforms the original vector of coefficients

$$f = (f_0, f_1, f_2, \ldots, f_{n-1})^t$$

into the new vector of values of $f()$, sampled at the above roots of unity:

$$\left(f(1), f(w), f\left(w^2\right), f\left(w^3\right), \ldots, f\left(w^{n-1}\right)\right)^t.$$

How to calculate this fast, in one go?

13.2 FFT: A Virtual Binary Tree

13.2.1 *FFT*

So, we need to sample our polynomial $f(z)$ at n distinct complex numbers:

$$z \in \left\{w, w^2, w^3, \ldots, w^n = 1\right\}.$$

How to do this fast? Break $f()$ into its even and odd parts:

$$f(z) = u(z^2) + z \cdot v(z^2),$$

where $u()$ and $v()$ are new polynomials, obtained from $f()$: $u()$ contains the even powers:

$$u(z) = f_0 + f_2 z + f_4 z^2 + f_6 z^3 + \cdots = \sum_{0 \le i < n, i \text{ is even}} f_i z^{i/2},$$

and $v()$ contains the odd powers:

$$v(z) = f_1 + f_3 z + f_5 z^2 + f_7 z^3 + \cdots = \sum_{0 \le i < n, i \text{ is odd}} f_i z^{(i-1)/2}.$$

What's so good about this new decomposition? Well, it is ready for recursion, to help sample both $u()$ and $v()$ at z^2.

13.2.2 *Virtual Binary Tree*

This works so long as n is even. Why? Because we can divide n by 2 and use recursion on $n/2$. To do this, look at our original root of unity: w. Raise it to power $n/2$:

$$w^{n/2} = \exp\left(\frac{2\pi\sqrt{-1}}{n} \cdot \frac{n}{2}\right) = \exp(\pi\sqrt{-1}) = -1.$$

This way, our original roots of unity split into two subsets — the upper ones, mirrored by the lower ones:

$$\begin{aligned} z \in & \{w, w^2, w^3, \ldots, w^{n/2-1}, w^{n/2}, w^{n/2+1}, w^{n/2+2}, \ldots, w^n = 1\} \\ = & \{w, w^2, w^3, \ldots, w^{n/2-1}\} \cup \{-1, -w, -w^2, -w^3, \ldots, 1\}. \end{aligned}$$

Here, the former $n/2$ points are on the upper semicircle in the complex plane. These are followed by $n/2$ more points that are only slightly different: they pick a minus sign and mirror underneath, on the lower semicircle (Figure 13.1). Each subset is now ready for recursion: it makes a subtree in a new (virtual) binary tree.

To do this, scan the points, one by one, and raise them to power 2. In the process, the minus sign drops, and both subsets get identical.

Thanks to this, both $u()$ and $v()$ should now be sampled on one and the same set, containing the same $n/2$ points:

$$z^2 \in \left\{w^2, w^4, w^6, \ldots, w^{n-2}, 1\right\}.$$

On this new set, how to sample $u()$ and $v()$? Recursively, of course! For this purpose, assume that n is a power of 2:

$$n = 2^k,$$

for some fixed integer $k \geq 0$. This way, recursion is indeed possible. Otherwise, extend f with a few tailing zeroes to make recursion possible.

Thanks to recursion, we can sample $u()$ and $v()$ fast as well. (This is the induction hypothesis, to be discussed soon.) This is FFT. Is it really fast? What is its complexity (total cost)?

13.2.3 *Complexity*

What is the total cost? Not too much: only $n \log_2 n$ complex multiplications and $n \log_2 n$ complex additions. To prove this, use mathematical induction on $k = 0, 1, 2, 3, \ldots$, or $n = 1, 2, 4, 8, \ldots$. For $n = 1$, there is no work at all. This agrees with the above formula:

$$1 \cdot \log_2 1 = 1 \cdot 0 = 0.$$

Now, what about $n = 2, 4, 8, 16, \ldots$? Assume that the above cost is correct for $n/2$. (This is the induction hypothesis.) This tells us the cost of each recursive call (to sample u or v at the above $n/2$ points):

$$\frac{n}{2} \log_2 \left(\frac{n}{2}\right) = \frac{n}{2} (\log_2 n - 1)$$

multiplications (and the same number of additions). On top of this, there is yet more work:

- n more multiplications to calculate z and $zv(z^2)$.
- On top of these, to add $u(z^2)$ and $zv(z^2)$, there are n more additions (or subtractions).

All these make a total of

$$2\frac{n}{2} (\log_2 n - 1) + n = n(\log_2 n - 1) + n = n \log_2 n$$

multiplications (and the same number of additions), as asserted.

13.3 The Inverse Transform

13.3.1 *Matrix Form*

The discrete Fourier transform can also be written in a matrix form:

$$f \to Wf,$$

where W is an $n \times n$ complex matrix, whose elements are roots of unity in their own right:

$$W \equiv \left(w^{ij}\right)_{0 \le i,j < n}.$$

This way, W is the same as its own transpose:

$$W = W^t.$$

13.3.2 *The Inverse Matrix*

Let's multiply W by its Hermitian adjoint. Since W has orthogonal columns (and orthogonal rows as well), we get

$$W^*W = WW^* = nI$$

(where I is the $n \times n$ identity matrix). This gives us the inverse of W:

$$W^{-1} = \frac{1}{n}W^* = \frac{1}{n}\bar{W}^t = \frac{1}{n}\bar{W}.$$

Let's use this to obtain the inverse transform easily.

13.3.3 *The Inverse Transform*

Thanks to this, the inverse transform looks like this: for a given n-dimensional vector g,

$$g \to W^{-1}g = \frac{1}{n}\bar{W}g.$$

This can be carried out in two stages:

- Apply FFT, with just one change: $\bar{w}$ rather than w.
- Divide by n.

Better yet, this can also be carried out in four stages:

- Take the complex conjugate of g. This produces $\bar{g}$.
- Apply FFT. This produces $W\bar{g}$.
- Take the complex conjugate of $W\bar{g}$. This produces $\bar{W}g$.
- Finally, divide by n.

13.4 Convolution

13.4.1 *Convolution*

Why is FFT so useful? Well, consider two polynomials: $f()$ and $g()$ (given explicitly, by their coefficients). How to calculate their product fg (explicitly, by its new coefficients)? Well, we could use a direct convolution, but this would be quite expensive. Better use FFT:

1. Introduce a few tailing zeroes, to make both f and g have $n = 2^k$ coefficients.
2. Use FFT to sample both $f()$ and $g()$ at the roots of unity (Figure 13.1).
3. Multiply the respective results, and obtain the new product fg at these points.
4. Use inverse FFT to transform back, and obtain the explicit coefficients of fg, as required.
5. We are not done yet. For example, if $f()$ and $g()$ are binary polynomials that represent long natural numbers, then we may have some more work to do. After all, fg may still contain coefficients (digits) that are not only 0 or 1 but also 2 or 3 or 4.... How to fix this? Push them ahead! For this purpose, start from the unit digit: the least significant digit. If it exceeds 1, then it must contribute to the more significant digit as well. For example, if it is 3, then reduce it to 1, and add 1 to the more significant digit. Then, advance to the more significant digit, and do the same, and so on.
6. Finally, drop leading zeroes, if any.

This is useful in cryptography. Still, this is not the only application: FFT is also useful in numerical analysis, signal processing, and more.

13.5 Exercises: Complexity

13.5.1 *FFT and Its Complexity*

1. Consider two long natural numbers. What is the best way to store them? *Hint*: as binary polynomials, with binary coefficients (digits): 0 or 1.
2. Multiply them, and obtain a new polynomial: their product (convolution), coefficient by coefficient.
3. How does this look like?
4. Is this a legitimate binary polynomial in its own right? *Hint*: not quite (see the following).
5. What is its length? *Hint*: like the sum of the lengths of the original natural numbers.
6. What coefficients (digits) are there in it? *Hint*: unfortunately, not only 0 or 1 but also 2 or 3 or 4....
7. How to fix this? *Hint*: push digits ahead (Section 13.4.1).
8. What is the extra cost of this? *Hint*: like the length of the product (the sum of the lengths of the original natural numbers).
9. Assume that the convolution uses a naive algorithm: a standard vertical multiplication (as in high-school). What is its cost (complexity)? *Hint*: $O(nm)$ bit operations (where n and m are the lengths of the original natural numbers).
10. Assume, for instance, that the original natural numbers contain as many as $n = m = 10^{10}$ binary digits. In this case, how big is the above cost? *Hint*: $nm = 10^{20}$.
11. To improve on this, use FFT (Sections 13.2.3–13.4.1). What is the cost now? Is it less than before? *Hint*: FFT costs $2n \log_2 n$ complex operations, each as expensive as 10^3 bit operations. This is still worthwhile:

$$10^3 \cdot 10 \cdot 10 \cdot 10^{10} \ll 10^{20}.$$

12. Implement FFT on your computer, and use it to carry out convolution.
13. Use this to multiply long natural numbers, stored as binary polynomials. In the end, don't forget to push digits ahead (Section 13.4.1). *Hint*: the solution can be found in Chapter 5 in Ref. [24].

Part V: Quantum Algorithms in Cryptography

What is the key ingredient in computer science? This is the bit: 0 or 1. Now, take k bits, one by one, in a row. What do you get? Well, each new bit doubles the number of possible configurations of 0's and 1's. So, in total, you get 2^k possible configurations. Each one makes a new binary index: $0 \le i < 2^k$. Each i indexes a node in a new (virtual) hypercube.

This will be the geometrical basis for our new quantum computer. For example, k bits will make a hypercube of dimension k, with 2^k binary nodes. How to index them? Take a configuration of k bits, and look at it as a binary index: $0 \le i < 2^k$.

In quantum mechanics, each bit is nondeterministic: a qubit. This way, the index i gets nondeterministic too: a new random variable. Each i indexes one node in the hypercube, at some probability. To calculate the probability, we need a (discrete) wave function. How? On the nodes, define a new (complex) grid function. This is not only geometrical but also algebraic: a new 2^k-dimensional vector, ready for all sorts of algebraic operations. Moreover, our new (discrete) wave function will be quite useful: it will tell us the corresponding probability. How? Square it up (and normalize), and you'll get the probability to have a particular i (and indeed a specific node in the hypercube).

What is so good about this? Well, to carry out an algebraic operation (or gate), no need to scan the nodes any more. Instead, do just one quantum operation. This way, in one go, you get to apply the same operation to the entire (discrete) wave function, as a whole. In parallel computing, this is called SIMD: single instruction, and multiple data.

This makes a new quantum computer, with a great computational power. After all, the total number of nodes grows exponentially with k. To use our quantum computer, we need a quantum algorithm, like quantum FFT [34].

Unfortunately, there is no quantum computer as yet: this is just science fiction. Still, it is a good idea to get ready. After all, technology progresses quite quickly. Besides, quantum computing also has a theoretical value: the logic behind it can help understand better both physics and computer science.

How to design a quantum algorithm? For example, look at FFT. It is defined algebraically, and can run recursively, even on a standard computer. But now, implement it on a (theoretical) quantum computer. This makes a much more efficient version: quantum FFT. It even leads to an efficient nonrecursive version. This is algebraic-geometrical again: the original recursion makes a (virtual) binary tree. Parse it, level by level, bottom to top, and you'll get a new nested loop, easy to run as you like: either on a traditional computer, or in parallel, or even on a quantum computer.

Thanks to quantum FFT, one could even break the RSA code in cryptography. Fortunately, there is nothing to worry about: quantum computers are not available yet. Still, it is worthwhile to learn how the code could break, and get ready, ahead of time.

Chapter 14

Quantum FFT: Recursive/Nonrecursive

What is a quantum computer? In information theory, the most elementary item is the bit: 0 or 1. How to design a yet higher structure? Take k bits, one by one, in a row. Together, they make a new geometrical structure: a (virtual) hypercube, made of $n = 2^k$ binary nodes. On it, you can now define a new (complex) grid function: $v \equiv v_i$ (indexed by $0 \le i < 2^k$, where i is made of our k original bits, and indexes one node in the hypercube).

This is our quantum computer. How simple! Its computational power grows exponentially with k (the number of bits). To implement an algorithm like FFT on it, use a (virtual) binary tree.

How to work in a binary tree? So far, we worked from top to bottom: recursive FFT. Here, on the other hand, we work the other way around: from bottom to top. This is more efficient. Virtually, this parses the entire binary tree in front of your eyes, level by level, bottom to top. No need to use a sophisticated (high-level) recursion any more. Instead, use an elementary (low-level) loop, to scan the nodes, level by level. This leads to nonrecursive FFT, easy to run on any computer: digital, parallel, or even quantum. Although there is no quantum computer as yet, it is a good idea to get prepared ahead of time, and write powerful quantum algorithms in advance.

14.1 Qubit and Its Gates

14.1.1 *Electron and Its Spin*

A bit is a variable with just two possible values: either 0 or 1. A qubit, on the other hand, is a random variable: it could be 0 at probability $|v_0|^2$, or 1 at probability $|v_1|^2 = 1 - |v_0|^2$. For example, the electron could spin up at probability $|v_0|^2$, or down at probability $|v_1|^2 = 1-|v_0|^2$. We'll never know for sure: after all, we're never going to look.

This is not the only way: there are many other ways to implement a qubit. Why use spin? Because, on the way, we'll get to learn a lot about spin, and how it behaves in a magnetic field (Figure 11.2). This will make a gate: an automatic mechanism to change v_0 and v_1 implicitly. This way, nature will serve as our new computer: it will store v_0 and v_1 for us and even manipulate them for us, implicitly and indirectly.

14.1.2 *Complex Degrees of Freedom*

In the digital computer, the bit is the smallest unit of memory (Section 6.11.3). Indeed, it stores just one bit: either 0 or 1. In our quantum computer, on the other hand, the electron stores much more: the entire state:

$$v \equiv \begin{pmatrix} v_0 \\ v_1 \end{pmatrix} \in \mathbb{C}^2,$$

containing two complex numbers: v_0 and v_1. Still, these complex variables are not quite independent of each other. After all, they must satisfy a normalization condition:

$$\|v\|^2 = v_0^2 + v_1^2 = 1.$$

Later on, the state will get much more complicated: v will get multidimensional and contain many more components. Only in the end will v get normalized. Until then, we can still work: use the components of v to store complex unknowns, even without knowing what they really are. To do this, let's start from our (two-dimensional) v and apply to it a 2×2 orthogonal matrix: a gate.

14.1.3 *An Orthogonal Pauli Matrix*

Our first gate is an orthogonal Pauli matrix:

$$\sigma_z \equiv \begin{pmatrix} 1 & \\ & -1 \end{pmatrix}$$

(Figure 14.1). (In the matrix, blank spaces stand for zero.) Let's apply it to v:

$$v \to \sigma_z v = \begin{pmatrix} v_0 \\ -v_1 \end{pmatrix}.$$

How to do this in practice? Let our electron fly rightward, through a magnetic field, oriented upward (Figures 11.9 and 11.10). Thanks to the Schrodinger equation, v_0 will then precess at frequency $-\omega$, and v_1 at frequency ω. Now, let's apply this from time $t = 0$ until time $t = t_0$, where

$$2\omega t_0 = \pi.$$

This way, v_1 will open a phaseshift of π (ahead of v_0). In other words, v_1 will pick a minus sign, as required. Unfortunately, there is a price

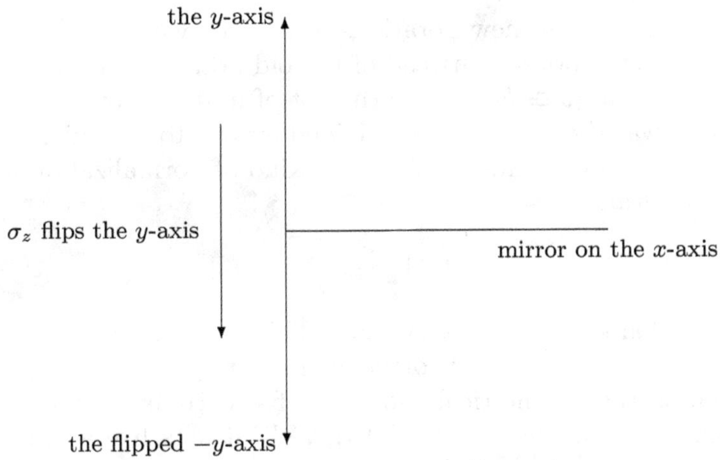

Figure 14.1. The Pauli matrix σ_z flips the y-axis. The x-axis acts like a mirror: on one hand, it remains fixed (eigenvalue: 1). On the other hand, it also has an active role: to help flip the y-axis (eigenvalue: -1).

to pay: an overall phaseshift of $-\pi/2$. This is global: it multiplies the entire vector v, as a whole. Physically, this has no effect at all. Numerically, on the other hand, this global factor should be stored elsewhere, until the end. Throughout the algorithm, it will change and pick more and more overall phaseshifts. Only in the end will it multiply v back again (as a whole).

14.1.4 *The Orthogonal Hadamard Matrix*

So far, we saw that the Pauli matrix σ_z acts like a mirror, placed on the (horizontal) x-axis. This marks the eigenvectors: the x-axis remains fixed (eigenvalue: 1), while the y-axis flips (eigenvalue: -1). The same principle works in yet another orthogonal matrix — the Hadamard matrix:

$$W_2 \equiv \frac{1}{\sqrt{2}} \begin{pmatrix} 1 & 1 \\ 1 & -1 \end{pmatrix}.$$

Is this familiar? This is actually the discrete Fourier transform, in the simple case of

$$k = 1, \quad n = 2^k = 2, \quad \text{and} \quad w = -1$$

(Figure 13.1). These new notations also agree with those in Figure 5.1, with a few updates. Instead of the old notations used there, use here $n \equiv N$ and $w \equiv F_{2,2}$ (the nth root of unity, before normalizing F). But F was then normalized (divided by $\sqrt{n}$, to get unitary). Here too, our W_n should undergo the same kind of normalization as well. In fact, we want

$$W_n \equiv F$$

(unitary, thanks to normalization). This is why our W_2 already contains factor $2^{-1/2}$, to get orthogonal, as required.

What is the geometrical meaning of the Hadamard transform, represented by our new matrix W_2? Well, it acts like a mirror too. But here, the mirror is oblique: it makes angle $\pi/8$ with the x-axis (Figure 14.2). (If you like, look at things the other way around: the x-axis makes angle $-\pi/8$ with the mirror.) Thanks to the mirror, the x-axis is mapped to the oblique line $y = x$ (that makes angle

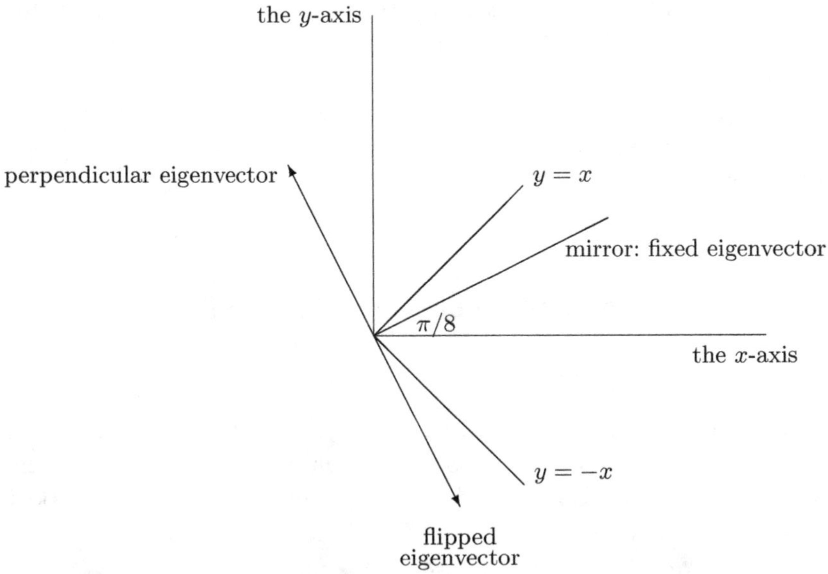

Figure 14.2. The Hadamard matrix acts like a mirror too: the x-axis is mapped to the oblique line $y = x$. The y-axis, on the other hand, is mapped to the oblique line $y = -x$. This way, the mirror remains fixed (at angle $\pi/8$), while its perpendicular picks a minus sign.

$\pi/8$ with the mirror but from the other side). This is just what we wanted:

$$\begin{pmatrix} 1 \\ 0 \end{pmatrix} \to W_2 \begin{pmatrix} 1 \\ 0 \end{pmatrix} = \frac{1}{\sqrt{2}} \begin{pmatrix} 1 \\ 1 \end{pmatrix}.$$

(This keeps the same norm: 1. Why? Because W_2 is orthogonal, thanks to the factor $2^{-1/2}$ in it.) The y-axis, on the other hand, makes angle $3\pi/8$ with the mirror. Once mirrored, it is mapped to the oblique line $y = -x$ (that makes angle $-3\pi/8$ with the mirror). This is just what we wanted:

$$\begin{pmatrix} 0 \\ 1 \end{pmatrix} \to W_2 \begin{pmatrix} 0 \\ 1 \end{pmatrix} = \frac{1}{\sqrt{2}} \begin{pmatrix} 1 \\ -1 \end{pmatrix}.$$

(Again, this preserves norm, as required.) Thus, the mirror marks the fixed eigenvector (of eigenvalue 1), and its perpendicular marks the flipping eigenvector (of eigenvalue -1).

How to apply W_2 in practice? Use a new magnetic field, oriented not upward but obliquely (at angle $\pi/4$ away from the vertical z-axis in 3-D). This way, the new magnetic field will point in between two axes (in 3-D): the vertical z-axis [represented by state $(1,0)^t$] and some horizontal axis that is equally likely to spin up or down [represented by state $(1,1)^t$]. This way, the new magnetic field will indeed align with the mirror, as required (Section 11.5.1).

14.2 Two Qubits and Their Grid Function

14.2.1 *Two Qubits*

Consider now a bigger system, containing two qubits: two electrons, each spinning up or down (at some probability). This makes a new (virtual) grid of four points, indexed by a new binary index: $i = 00, 01$ (top row in Figure 14.3) and $i = 10, 11$ (bottom row).

14.2.2 *Binary Index: A New Random Variable*

Here, i is a random variable in its own right. In fact, i could take either of these four values, each at some probability. More specifically, $i = 0$ (or 1 or 2 or 3), if the electrons spin up-up (or up-down or down-up or down-down), which may happen at probability $|v_{00}|^2$ (or $|v_{01}|^2$ or $|v_{10}|^2$ or $|v_{11}|^2$, respectively). This way, the state v is now

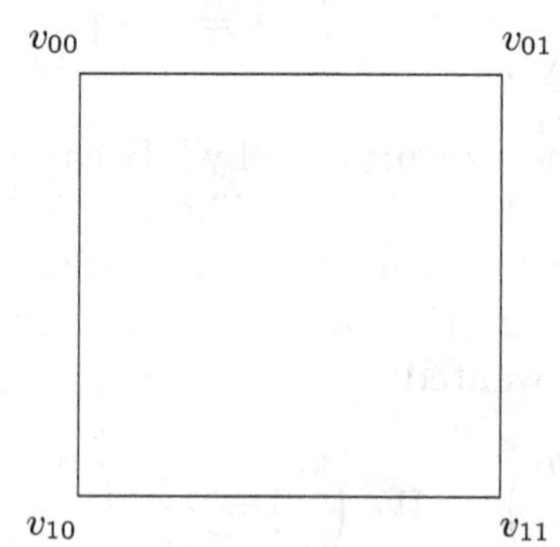

Figure 14.3. Two (entangled) qubits make a (virtual) grid of four points, indexed (row by row) by the binary index $i = 00, 01, 10, 11$. On this grid, define the (normalized) grid function $v \equiv v_i$. How likely is the first electron to spin up (or 0), while the second spins down (or 1)? The probability for this is $|v_{01}|^2$.

interpreted not only algebraically (as a vector of norm 1) but also geometrically (as a new grid function).

14.2.3 *Entanglement: Nonseparable Grid Function*

This is indeed entanglement: the qubits depend on each other and can never disregard one another. In other words, the values of v on the top row are not proportional to those on the bottom row. Likewise, the values of v on the left column are not proportional to those on the right column in the grid. Why? Because v is nonseparable (Section 8.3.2).

14.2.4 *Two Qubits and Their Gate*

Why is this efficient? Because you can now apply a gate (a 2×2 unitary matrix) to the first qubit. As a result, the same gate will act on both columns at the same time. Likewise, if you like, apply any gate to the second qubit. As a result, this gate will act on both rows at the same time. Either way, you benefit: in just one go, you apply the same gate twice — either to both columns or to both rows. Either way, you apply your gate to the entire grid function v.

In parallel computing, this is called SIMD: a single instruction, carried out on multiple data. This is why quantum computing is so powerful: it could carry out your algorithm in parallel, without paying any price!

14.2.5 *An Odd-Even Ordering*

So far, our grid was ordered row by row: the top row, followed by the bottom row. But we could also order it column by column: the left column (indexed by an even index: $i = 00, 10$), followed by the right column (indexed by an odd index: $i = 01, 11$). This is the odd-even ordering. There is no physical change: the first qubit is still first, and the second second. Only priority changes: the second qubit gets prior!

In this new ordering, the second qubit is like an ugly duckling: it was less significant and is now more significant! We are now ready to apply FFT to our v. For this purpose, set

$$k = 2, \quad n = 2^k = 4, \quad \text{and} \quad w = \sqrt{-1}$$

(the fourth root of unity). What phase should w take? Well, it depends: in the Fourier transform, positive. In its inverse, on the other hand, negative (or the other way around).

14.3 FFT on Two Qubits

14.3.1 *Unitary FFT*

Thanks to the new odd-even ordering, FFT takes the form of a triple product of three 4×4 (unitary) matrices: a Hadamard matrix, followed by a diagonal matrix, followed by another Hadamard matrix. To have a unitary FFT, we must divide by $\sqrt{2}$ once again. Together with the old $\sqrt{2}$ in W_2, this will make $\sqrt{4}$, as required:

$$W_4 \equiv \frac{1}{\sqrt{2}} \begin{pmatrix} I & I \\ I & -I \end{pmatrix} \begin{pmatrix} 1 & & & \\ & 1 & & \\ & & 1 & \\ & & & w \end{pmatrix} \begin{pmatrix} W_2 & \\ & W_2 \end{pmatrix}$$

(where I is the 2×2 identity matrix). Let's look into this.

14.3.2 *Hadamard Transform on the First Qubit*

What do we have here? This is a triple product of three 4×4 (unitary) matrices. To apply it to v, work from right to left. First, apply the rightmost matrix. This is W_2: the Hadamard transform, on each column in the grid. (Thanks to the odd-even ordering, the grid is already ordered column by column, not row by row.) To implement this, apply W_2 to the first qubit only. After all, this qubit stores the values of v in each individual column. Thus, W_2 on the first qubit is the same as W_2 on both columns at the same time, as required.

14.3.3 *Second Qubit: Control Qubit*

At the middle of our triple product, we have a diagonal unitary matrix. Let's call it D_1:

$$D_1 \equiv \begin{pmatrix} 1 & & & \\ & 1 & & \\ & & 1 & \\ & & & w \end{pmatrix},$$

where

$$w = \sqrt{-1}.$$

On its main diagonal, D_1 has 1's, except for the lower-right element, which is imaginary: $w = \sqrt{-1}$ (the fourth root of unity). How to apply D_1 to a vector like v? For this, factorize D_1 as a product (of two other diagonal matrices):

$$D_1 = \begin{pmatrix} 1 & & & \\ & 1 & & \\ & & w^{-1/2} & \\ & & & \sqrt{w} \end{pmatrix} \begin{pmatrix} 1 & & & \\ & 1 & & \\ & & \sqrt{w} & \\ & & & \sqrt{w} \end{pmatrix}.$$

How to apply this to a vector like v? Again, work from right to left: apply the right matrix, followed by the left one. This makes two stages:

- The second qubit tells us in what column we are in the grid. In other words, it stores each individual row: the top and the bottom too. To this qubit, apply the gate in Section 14.1.3, only for a new (shorter) time t_0, for which

  $$2\omega t_0 = \frac{\pi}{4}.$$

  This opens a new phaseshift: the right column gets ahead of the left one. In other words, the third and fourth components of v (in its odd-even ordering) are now at phase $\pi/4$ ahead.
- But this is not enough. We still need to phaseshift them even more: the fourth component by $\pi/4$ more and the third by $\pi/4$ back again. In total, this will phaseshift the fourth by $\pi/2$, and the third by none, as required. How to do this?

1. First, nail the first and second components and make sure they stay put. How to do this?
2. For this, we need a mechanism to find out in which column we are (without changing the state v). How to design this?
3. Use the second qubit as a control qubit: assume that our quantum computer has an inner detector, to measure the spin of the second qubit, and use the answer immediately, and destroy it soon after, never recording it, and never telling us what it was. This keeps v as is.
4. Only our quantum computer knows the answer and can now use it to decide what to do:
   - If the second qubit spins up (has value 0), then we are in the left column. In this case, do nothing. After all, the first and second main-diagonal elements are both 1.
   - If, on the other hand, the second qubit spins down (has value 1), then we are in the right column (containing the third and fourth components, in their odd-even ordering). In this case, use our magnetic field to apply the gate

     $$\begin{pmatrix} 1 & \\ & w \end{pmatrix}$$

     to the first qubit. For this, pick a new t_0 (twice as long) so that

     $$2\omega t_0 = \frac{\pi}{2}.$$

     This will only change the third and fourth components (the right column, in its odd-even ordering): the fourth component will gain phase $\pi/4$ (to make a total gain of $\pi/2$ since the start), as required. The third, on the other hand, will lose phase $\pi/4$, to fit with the first and second components back again, as required.

So far, the second qubit was in charge: it had the upper hand and told our quantum computer what to do with the first qubit. But this is not a must: we could also work the other way around and interchange their roles. After all, our task is completely symmetric: D_1 doesn't distinguish between the qubits in any way.

14.3.4 Hadamard Transform on the Second Qubit

But don't forget: in our triple product, we still have the leftmost matrix to apply. This is actually the Hadamard matrix W_2, applied to each individual row in the grid (in its odd-even ordering). To do this, apply W_2 to the second qubit only. After all, this qubit stores the values of v in each individual row. Thus, applying W_2 to it actually applies W_2 twice for free: to both rows at the same time, as required. This is the power of quantum computing.

14.4 Three Qubits: Discrete Cube

14.4.1 Three Entangled Qubits

Next, let's move on to a yet higher dimension:

$$k = 3, \quad n = 2^k = 8, \quad \text{and} \quad w = (-1)^{1/4}$$

(the eighth root of unity). This way, we have not only two but three (entangled) qubits. Between them, they make a new random variable: the new binary index

$$i = 000, 001, 010, 011, 100, 101, 110, 111$$

(Figure 14.4). These have a new geometrical meaning: corners of the unit cube. On them, we can now define a new grid function (eight-dimensional complex vector, of norm 1):

$$v \equiv v_i.$$

This way, $|v_i|^2$ is the probability to have index i. For example, $|v_{110}|^2$ is the probability of down-down-up: two spin-down electrons, followed by a spin-up electron. (For numerical purposes, we also need to store an overall complex coefficient.)

This is our new quantum computer, with its great power: in just three qubits, nature stores eight (complex) degrees of freedom for us (minus one, due to normalization). To our new v, let's apply our unitary FFT.

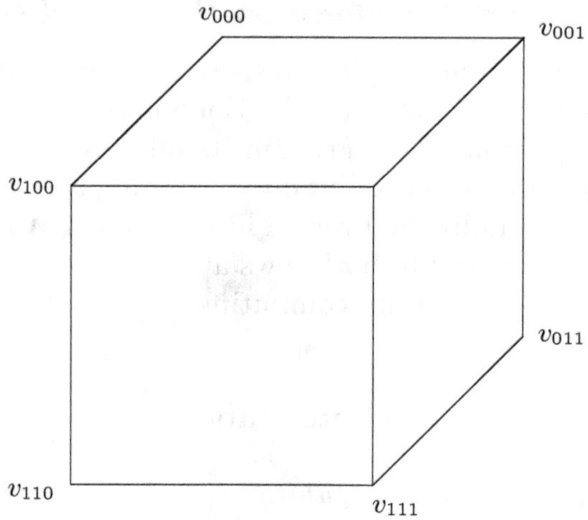

Figure 14.4. Three qubits make a cube of eight points, indexed by the binary index $i = 000, 001, 010, 011, 100, 101, 110$, and 111, with the (normalized) grid function $v \equiv v_i$. How likely is up-down-down? The probability for this is $|v_{011}|^2$.

14.4.2 *Unitary FFT*

To apply our unitary FFT, let's reorder our cube, using a new odd-even ordering: start from the left part of the cube (indexed by an even i), followed by the right part (indexed by an odd i). There is no physical change: the first qubit is still first, and the last last. Only priority changes: the third qubit is now our new "ugly duckling": it was least prior and is now most prior! Thanks to this, FFT takes the form of a new triple product (of three 8×8 unitary matrices):

$$W_8 \equiv \frac{1}{\sqrt{2}} \begin{pmatrix} I & I \\ I & -I \end{pmatrix} \begin{pmatrix} I & & & & \\ & 1 & & & \\ & & w & & \\ & & & w^2 & \\ & & & & w^3 \end{pmatrix} \begin{pmatrix} W_4 & \\ & W_4 \end{pmatrix}$$

(where I is now the 4×4 identity matrix). How to apply this to v?

14.4.3 *Recursion*

How to apply this triple product to a vector like v? Work from right to left: start from the rightmost matrix. How to apply it? Call FFT recursively, on both parts: the left part (even i's) and the right part (odd i's), at the same time. How to do this? Leave the "ugly duckling" out of the game: whatever it may be (0 or 1), W_4 will act in the same way (on both the left and right parts), as required. This is the power of quantum computing: get parallel, for free! (Really free: no hidden charge....)

Thus, in the recursive call, the ugly duckling remains idle. The two other qubits, on the other hand, are active: W_4 acts on them. This will take care of both parts at the same time, as required. Logically, this will reduce the original binary tree into a (thinner) unary tree, saving a lot of work.

14.4.4 *Diagonal Unitary Matrix*

In the latter triple product, at the middle, we have an 8×8 diagonal (unitary) matrix. On its main diagonal, in the lower-right block, it has powers of w:

$$w = (-1)^{1/4}, \quad w^2 = \sqrt{-1}, \quad \text{and} \quad w^3 = (-1)^{3/4}.$$

(What is their phase? In the Fourier transform — positive. In its inverse — negative.) Is this familiar? This new w^2 was called w in Section 14.3.3 and was used there to define D_1. Here, we'll apply D_1 to a pair of qubits: the ugly duckling and one more qubit. For this, factorize our diagonal matrix as a product of two:

$$\begin{pmatrix} I &&&& \\ & 1 &&& \\ && w && \\ &&& w^2 & \\ &&&& w^3 \end{pmatrix} = \begin{pmatrix} I &&&& \\ & 1 &&& \\ && 1 && \\ &&& w^2 & \\ &&&& w^2 \end{pmatrix} \begin{pmatrix} I &&&& \\ & 1 &&& \\ && w && \\ &&& 1 & \\ &&&& w \end{pmatrix}.$$

How to apply this? Say, from right to left: start from the right matrix. How to apply it? Pick a pair of two active qubits:

1. Our ugly duckling.
2. From the other two, pick the inferior.

To this pair, apply D_2: same as D_1, except for one change: t_0 twice as short, to help design our new w. This will do the trick.

And how to apply the left matrix? Pick another pair:

1. Our ugly duckling.
2. From the other two, pick now the superior (rather than the inferior).

To this new pair, apply the original D_1 (with the old w, called here w^2). This will do the trick, as required.

So, how are D_1 and D_2 different? Well, D_1 is higher (applied to a higher pair, of a coarse scale). To fit in this, it must use the old w (called here w^2). D_2, on the other hand, is lower (applied to a lower pair, of a fine scale). To fit in this, it uses our new w (with its small phase).

Still, they also have something in common: both are diagonal and commute — it doesn't matter which comes first and which follows. Later on, we'll extend this even more and design D_3, $D_4, \ldots$, and so on. Before going into this, let's finish our FFT on the cube.

14.4.5 *Final Hadamard Transform*

Finally, apply the Hadamard transform W_2 to the "ugly duckling." This is most efficient: in our discrete cube, W_2 will act on four rows at the same time. This is indeed SIMD: the same operation is carried out on all rows, in parallel. Why? Because the two other qubits remain idle (inactive) and only help duplicate (implicitly) on all rows, for no more cost! This completes our FFT on the cube. Let's extend this to a yet higher dimension: a hypercube.

14.5 *k* Qubits: Hypercube

14.5.1 *k Entangled Qubits*

Let's move on to a yet higher dimension, using a yet larger k:

$$k > 3, \quad n = 2^k, \quad \text{and} \quad w \equiv (-1)^{2/n}$$

(the nth root of unity). This way, n grows exponentially, and w decreases exponentially (in phase): as k grows by one, n doubles, and w halves (in phase) and becomes $\sqrt{w}$. Fortunately, w can remain implicit: never stored or calculated explicitly at all! As a matter of fact, as we'll see in the following, w is only introduced to help clarify the presentation.

14.5.2 *Hypercube*

Assume now that our quantum computer contains $k > 3$ (entangled) qubits. This way, v will be much bigger than before: now grid function, defined on a hypercube of $n = 2^k$ nodes. This is the great power of quantum computing: in just k qubits, nature will store as many as $n = 2^k$ degrees of freedom for us (minus one, due to normalization). (For numerical purposes, we also need to store an external complex coefficient, to accumulate global phaseshifts of v, as a whole.)

So far, we introduced FFT for $k \le 3$ and $n \le 8$. We are now ready to extend this to the general case as well. This will be done by mathematical induction on $k = 1, 2, 3, \ldots$. On the way, we'll also get rid of recursion altogether and parse it in an efficient loop.

14.5.3 *Unitary FFT*

For $k \le 3$, we already wrote FFT as a triple product of three (unitary) matrices. Let's extend this to a more general k, with

$$n \equiv 2^k \quad \text{and} \quad w \equiv (-1)^{2/n}$$

(the nth root of unity). In this case, FFT makes a new triple product:

$$W_n = \frac{1}{\sqrt{2}} \begin{pmatrix} I & I \\ I & -I \end{pmatrix} \begin{pmatrix} I & & & & & & & \\ & 1 & & & & & & \\ & & w & & & & & \\ & & & w^2 & & & & \\ & & & & w^3 & & & \\ & & & & & \ddots & & \\ & & & & & & w^{n/2-2} & \\ & & & & & & & w^{n/2-1} \end{pmatrix}$$

$$\times \begin{pmatrix} W_{n/2} & \\ & W_{n/2} \end{pmatrix},$$

where I is the $(n/2) \times (n/2)$ identity matrix.

14.5.4 *The Odd-Even Ordering*

In the leftmost matrix, I appears four times, in four blocks. What do these I's stand for? Well, in the odd-even ordering, the kth qubit is an "ugly duckling:" it was least significant and is now most significant, prior to all others! Thus, these four I's tell us the state of the ugly duckling: spin-up (as in the upper-left block) or spin-down (as in the lower-right block). This is the only active qubit: all others are left out of the game: completely idle. Their job is just to duplicate and carry out the same operation over and over again, as in SIMD. We'll come back to this later.

14.6 Recursive vs. Nonrecursive FFT

14.6.1 *Recursion*

How to apply our triple product to v? Start from the right: apply the rightmost matrix: recursive FFT (on the $k-1$ leading qubits). This leaves the kth qubit idle (to help duplicate). Who is this? This is our ugly duckling! After all, in the odd-even ordering, nothing changed physically. In particular, the ugly duckling didn't move at all: it is still last but with top priority now. This way, FFT acts (recursively) twice at the same time: not only on the first block (even-numbered components of v, where the ugly duckling spins up) but also on the second subvector (odd-numbered ones, where the ugly duckling spins down). Thanks to quantum computing, this is done simultaneously, in parallel, in just one recursive call.

14.6.2 *Recursion and Mathematical Induction*

Recursion mirrors mathematical induction. Later on, we'll use mathematical induction to design a nonrecursive version too and prove its correctness. In the induction step, we'll advance from $k-1$ to k. In the process, what happens to the dimension? It doubles: from $n/2$ to n. In the process, what happens to the root of unity? It halves (in phase):

$$w \leftarrow \sqrt{w}.$$

Indeed, the old root of unity was defined from $n/2$ (rather than n):

$$(-1)^{2/(n/2)} = (-1)^{4/n} = w^2$$

(where w is already new). Let's use the old root of unity in the induction hypothesis.

14.6.3 *Diagonal Unitary Matrix*

In our triple product, at the middle, we have now an $n \times n$ diagonal (unitary) matrix. How to apply it to our up-to-date v? Use mathematical induction on $k = 3, 4, 5, \ldots$:

1. Assume that we already have $D_1, D_2, \ldots, D_{k-2}$. (This is the induction hypothesis.)
2. Apply D_1 to the first and kth qubits, D_2 to the second and kth qubits, ..., and D_{k-2} to the $(k-2)$th and kth qubits.
3. Don't forget: our ugly duckling never moved — it is still the kth qubit. Only its priority changed: in our odd-even ordering, it now has top priority.
4. In the above process, only one qubit was inactive (idle): the $(k-1)$th qubit (of least priority). What for? To help duplicate (on the finest scale).
5. Thus, the above process already produced the diagonal matrix

$$\begin{pmatrix} I & & & & & & & \\ & 1 & & & & & & \\ & & 1 & & & & & \\ & & & w^2 & & & & \\ & & & & w^2 & & & \\ & & & & & \ddots & & \\ & & & & & & w^{n/2-2} & \\ & & & & & & & w^{n/2-2} \end{pmatrix}.$$

6. This was thanks to the induction hypothesis. In it, w^2 and $n/4$ played the role of w and $n/2$, respectively.
7. But this diagonal matrix is not quite what we want. How to fix it?

8. For this, define D_{k-1}: same as D_{k-2}, except for one change: our new w (rather than w^2).
9. Once D_{k-1} is ready, apply it to the $(k-1)$th and kth qubits. This produces yet another diagonal matrix:

$$\begin{pmatrix} I & & & & & & & \\ & 1 & & & & & & \\ & & w & & & & & \\ & & & 1 & & & & \\ & & & & w & & & \\ & & & & & \ddots & & \\ & & & & & & 1 & \\ & & & & & & & w \end{pmatrix}.$$

10. Here, things are the other way around: the $k-2$ leading qubits remain idle and only help duplicate (at all intermediate scales).
11. Finally, look at these two (diagonal) matrices. Their product is

$$\begin{pmatrix} I & & & & & & & \\ & 1 & & & & & & \\ & & w & & & & & \\ & & & w^2 & & & & \\ & & & & w^3 & & & \\ & & & & & \ddots & & \\ & & & & & & w^{n/2-2} & \\ & & & & & & & w^{n/2-1} \end{pmatrix}.$$

12. This is the diagonal matrix we want. It was just applied to v, as required.

14.6.4 *Final Hadamard Transform*

Finally, apply the Hadamard transform to the kth qubit. This activates the ugly duckling (that distinguishes odd from even and makes the coarsest scale). The $k-1$ leading qubits, on the other hand, remain idle and only help duplicate (on all finer scales). This completes our (recursive) FFT. Still, recursion can be rather expensive. Is there a better version?

14.6.5 *Nonrecursive Version: Nested Loops*

How to avoid recursion altogether and parse it in a (nested) loop? Use a new algorithm:

- Scan the qubits, one by one, as follows: for $j = 1, 2, \ldots, k$, look at the jth qubit. This will be our ugly duckling (starting from the innermost recursive call, back out). Do the following:
  - If $j = 1$, then the next loop is empty and can be skipped.
  - Use an inner loop to scan pairs of qubits, as follows: apply D_1 to the first and jth qubits, D_2 to the second and jth qubits,..., and D_{j-1} to the $(j-1)$th and jth qubits.
  - At the end of this inner loop, apply the Hadamard transform W_2 to the jth qubit.

Why is this correct? To prove this, use mathematical induction on $k = 1, 2, 3, \ldots$. For $k = 1$, this is easy: the inner loop is empty, and our algorithm just applies W_2, as required. Now, for $k \geq 2$, assume that the algorithm works for $k - 1$. (This is the induction hypothesis.) Now, look at the subloop $j = 1, 2, \ldots, k - 1$. Thanks to the induction hypothesis, this actually calls FFT (recursively), on the $k-1$ leading qubits, as required. Only the ugly duckling remains idle, to help duplicate (on the coarsest scale, which was finest in the original ordering). This way, this is done twice: for not only even- but also odd-numbered components of v, as required.

We are not done yet. Now, the outer loop reaches the final qubit: $j = k$. There, the inner loop runs for the last time. This applies our diagonal matrix (Section 14.6.3). Finally, W_2 is applied to the kth qubit, as required.

This is a nested loop over k qubits. Is this expensive? Not at all: just $O(k^2)$. This is quite fast: much faster than any digital computer. Can you build a quantum computer for us?

14.7 Exercises: Larmor Precession

14.7.1 *Electron and Its Spin*

1. Let's model an electron and its spin.
2. For this, assume that our detector points at

$$u \equiv \begin{pmatrix} u_1 \\ u_2 \\ u_3 \end{pmatrix} \in \mathbb{R}^3, \quad \|u\| = 1.$$

3. How likely is an electron to spin (counterclockwise) around u? *Hint*: the probability for this is $\cos^2(\theta/2)$, where θ is the angle between u and the spin vector of the electron.
4. Look at a simple example, in which these vectors are perpendicular:

$$\theta = \frac{\pi}{2}.$$

5. What does this mean physically? *Hint*: the electron has a neutral state — it is equally likely to spin clockwise or counterclockwise around u.
6. Does this agree with our formula? *Hint*: yes:

$$\cos^2\left(\frac{\theta}{2}\right) = \cos^2\left(\frac{\pi}{4}\right) = \frac{1}{2}.$$

14.7.2 *Magnetic Field and Spin*

1. Consider a (uniform) magnetic field, pointing upward, in the positive z-direction (Figure 14.5):

$$\mathbf{B} \equiv \|\mathbf{B}\|u, \quad \text{where} \quad u \equiv \begin{pmatrix} 0 \\ 0 \\ 1 \end{pmatrix}.$$

2. In this magnetic field, let's guess a dynamic (time-dependent) state for the spin of our electron:

$$v \equiv v(t) \equiv \begin{pmatrix} v_0(t) \\ v_1(t) \end{pmatrix} \equiv \begin{pmatrix} \cos\left(\frac{\theta}{2}\right)\exp(-i\phi t) \\ \\ \sin\left(\frac{\theta}{2}\right)\exp(i\phi t) \end{pmatrix},$$

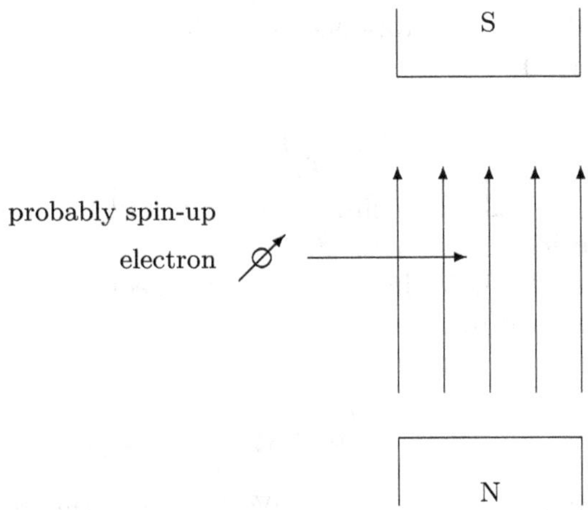

Figure 14.5. The uniform magnetic field points upward: $\mathbf{B} \equiv \|\mathbf{B}\|(0,0,1)^t$.

where $i = \sqrt{-1}$ (the imaginary number) and θ is the (unknown) angle between the spin vector and the vertical z-axis.

3. What is v? *Hint*: v is the state of the electron, in terms of spin-up-or-down.
4. Give an example. *Hint*: $v = (1,0)^t$ is deterministic — spin-up ($\theta = 0$).
5. Give yet another example. *Hint*: $v = (0,1)^t$ is deterministic too — spin-down ($\theta = \pi$).
6. In general, is v physical? *Hint*: v is more mathematical than physical. Indeed, v is only two-dimensional, not three-dimensional. Besides, v is complex, not real.
7. Look at the dynamic state $v(t)$ again. Is it just a guess? *Hint*: it is much more — it solves the Schrodinger equation (with Hamiltonian proportional to σ_z).
8. In $v(t)$, ϕ is the Larmor frequency (proportional to $\|\mathbf{B}\|$).
9. What is $\|v\|$? *Hint*: $\|v\| = 1$, as required.
10. What is the phaseshift between v_0 and v_1? *Hint*: $2\phi t$.
11. At what time is the phaseshift exactly π? *Hint*: at time $\pi/(2\phi)$.
12. By this time, what happened to v? *Hint*: v_1 was multiplied by i and v_0 by $-i$.
13. Up to an overall phaseshift, what happened to v? *Hint*: v_1 picked a minus sign.

14. Algebraically, what happened to v? *Hint*: v was multiplied by σ_z (from the left):

$$v\left(\frac{\pi}{2\phi}\right) = \sigma_z v(0)$$

(up to an overall phaseshift, which should be stored elsewhere, for a numerical purpose).

15. In our magnetic field, how likely is the electron to spin up (counterclockwise around the z-axis)? *Hint*: the relevant observable is a diagonal 2×2 matrix:

$$\tilde{S}_z = \frac{\bar{h}}{2}\sigma_z = \frac{\bar{h}}{2}\begin{pmatrix} 1 & 0 \\ 0 & -1 \end{pmatrix}.$$

Its relevant eigenvector is $(1,0)^t$. To have the required probability, look at this eigenvector. Take its inner product with v, calculate the absolute value, and square it up:

$$|(1,0)v|^2 = |v_0|^2 = \left|\cos\left(\frac{\theta}{2}\right)\right|^2 = \cos^2\left(\frac{\theta}{2}\right).$$

14.7.3 *Spin-Up Observable and Its Expectation*

1. What is an observable? *Hint*: this is just another word for measurable. Algebraically, this is a Hermitian matrix, to help observe (or measure) a random variable.
2. Look at the spin-up random variable. What is its expectation at v? *Hint*: the spin-up observable is $\tilde{S}_z = \bar{h}\sigma_z/2$. Its expectation at v is

$$\begin{aligned}
\left(v, \tilde{S}_z v\right) &= v^* \tilde{S}_z v \\
&= \frac{\bar{h}}{2}\bar{v}^t \sigma_z v \\
&= \frac{\bar{h}}{2}\left(\cos\left(\frac{\theta}{2}\right)\exp(i\phi t), \sin\left(\frac{\theta}{2}\right)\exp(-i\phi t)\right) \\
&\quad \times \begin{pmatrix} 1 & 0 \\ 0 & -1 \end{pmatrix}\begin{pmatrix} \cos\left(\frac{\theta}{2}\right)\exp(-i\phi t) \\ \sin\left(\frac{\theta}{2}\right)\exp(i\phi t) \end{pmatrix} \\
&= \frac{\bar{h}}{2}\left(\cos^2\left(\frac{\theta}{2}\right) - \sin^2\left(\frac{\theta}{2}\right)\right) \\
&= \frac{\bar{h}}{2}\cos(\theta).
\end{aligned}$$

14.7.4 *Spin-Right Observable*

1. Next, look at another random variable: spin-right (counterclockwise around the x-axis in 3-D). What is its expectation at v? *Hint*: the relevant observable is now

$$\tilde{S}_x = \frac{\bar{h}}{2}\sigma_x = \frac{\bar{h}}{2}\begin{pmatrix} 0 & 1 \\ 1 & 0 \end{pmatrix}.$$

At v, its expectation is

$$\begin{aligned}
\left(v, \tilde{S}_x v\right) &= v^* \tilde{S}_x v \\
&= \frac{\bar{h}}{2}\bar{v}^t \sigma_x v \\
&= \frac{\bar{h}}{2}\left(\cos\left(\frac{\theta}{2}\right)\exp(i\phi t), \sin\left(\frac{\theta}{2}\right)\exp(-i\phi t)\right)\begin{pmatrix} 0 & 1 \\ 1 & 0 \end{pmatrix} \\
&\quad \times \begin{pmatrix} \cos\left(\frac{\theta}{2}\right)\exp(-i\phi t) \\ \sin\left(\frac{\theta}{2}\right)\exp(i\phi t) \end{pmatrix} \\
&= \frac{\bar{h}}{2}\left(\cos\left(\frac{\theta}{2}\right)\sin\left(\frac{\theta}{2}\right)\exp(2i\phi t)\right. \\
&\quad \left. + \sin\left(\frac{\theta}{2}\right)\cos\left(\frac{\theta}{2}\right)\exp(-2i\phi t)\right) \\
&= \frac{\bar{h}}{2} 2\sin\left(\frac{\theta}{2}\right)\cos\left(\frac{\theta}{2}\right)\frac{\exp(2i\phi t) + \exp(-2i\phi t)}{2} \\
&= \frac{\bar{h}}{2}\sin(\theta)\cos(2\phi t).
\end{aligned}$$

14.7.5 *Spin-In Observable*

1. Next, look at yet another random variable: spin-in (counterclockwise around the y-axis in 3-D). What is its expectation at v? *Hint*: the relevant observable is now

$$\tilde{S}_y = \frac{\bar{h}}{2}\sigma_y = \frac{\bar{h}}{2}\begin{pmatrix} 0 & -i \\ i & 0 \end{pmatrix}.$$

At v, its expectation is

$$\begin{aligned}
\left(v, \tilde{S}_y v\right) &= v^* \tilde{S}_y v \\
&= \frac{\bar{h}}{2} \bar{v}^t \sigma_y v \\
&= \frac{\bar{h}}{2} \left(\cos\left(\frac{\theta}{2}\right) \exp(i\phi t), \sin\left(\frac{\theta}{2}\right) \exp(-i\phi t) \right) \\
&\quad \times \begin{pmatrix} 0 & -i \\ i & 0 \end{pmatrix} \begin{pmatrix} \cos\left(\frac{\theta}{2}\right) \exp(-i\phi t) \\ \sin\left(\frac{\theta}{2}\right) \exp(i\phi t) \end{pmatrix} \\
&= \frac{\bar{h}}{2} \left(\cos\left(\frac{\theta}{2}\right) \sin\left(\frac{\theta}{2}\right) (-i) \exp(2i\phi t) \right. \\
&\quad \left. + \sin\left(\frac{\theta}{2}\right) \cos\left(\frac{\theta}{2}\right) i \exp(-2i\phi t) \right) \\
&= \frac{\bar{h}}{2} 2 \sin\left(\frac{\theta}{2}\right) \cos\left(\frac{\theta}{2}\right) \frac{\exp(2i\phi t) - \exp(-2i\phi t)}{2i} \\
&= \frac{\bar{h}}{2} \sin(\theta) \sin(2\phi t).
\end{aligned}$$

2. What is the mathematical meaning of $2\phi t$? *Hint*: the phaseshift between v_0 and v_1.

14.7.6 *Expected Spin Vector*

1. By now, we have three expectations. Place them in a new column vector in 3-D:

$$\frac{\bar{h}}{2} \left(\sin(\theta) \cos(2\phi t), \sin(\theta) \sin(2\phi t), \cos(\theta)\right)^t .$$

   Hint: see Figure 14.6.
2. How does this vector look like? *Hint*: at frequency 2ϕ, it rotates around the vertical z-axis (at angle θ away, in 3-D).
3. This is Larmor's precession.
4. Does it model any classical phenomenon as well? *Hint*: this is also how a charged sphere would spin classically.

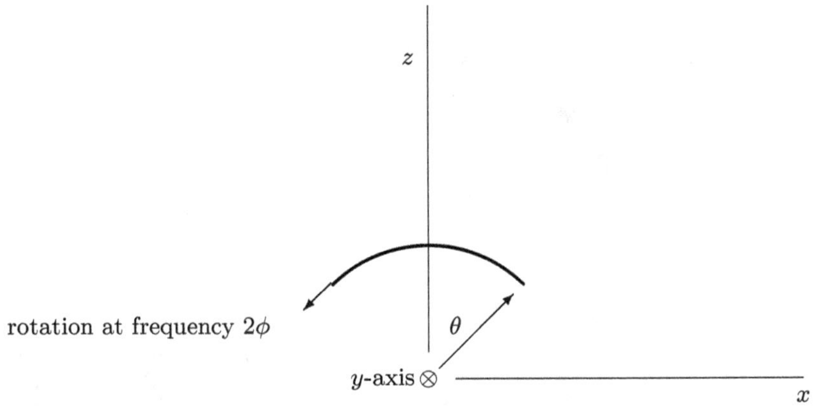

Figure 14.6. Larmor precession: a view from the side. The expected spin vector precesses around the (vertical) z-axis (at frequency 2ϕ).

5. Why? *Hint*: this is a special case of a more general theorem — in quantum mechanics, the expectations obey the classical law.
6. Simulate quantum FFT on your digital computer.
7. Does it work?
8. How to confirm this? *Hint*: apply inverse quantum FFT. Do you get the original v back again?

Chapter 15

Shor's Factoring Algorithm

What is a quantum computer? Mathematically, it is quite simple: just a list of k qubits. Each qubit is a random variable in its own right: either 0 or 1. Still, they are entangled: their wave function can change with one another. How? If in phase only, then this is not a real change: they are still independent (Section 6.8.3). If in magnitude too, then this is already a real change. In this case, they really depend on one another: if one switches from 0 to 1, then the others change distribution.

Either way, at some (joint) probability, each qubit could spin up (value: 0) or down (value: 1). This makes a (virtual) hypercube of 2^k nodes. On it, define a (complex) grid function: the joint wave function. Its square is the probability to have a particular configuration of 0's and 1's. There are as many as 2^k different configurations, each at one node. At this node, look at the (normalized) joint wave function, take its absolute value, and square it up, and you'll get the probability to have this particular configuration. This is the great power of quantum computing: with as little as k qubits, you get as many as $2^k - 1$ (complex) degrees of freedom (the joint wave function, at the nodes).

Unfortunately, nobody built a real quantum computer as yet. By now, this is still science fiction. Still, better get ready. After all, technology progresses quite fast. You may wake up one morning and find a new quantum computer on your desk...

Still, nothing can stop us from "playing" with this in our imagination, and invent all sorts of quantum algorithms that may be quite practical one day. This may give us not only a lot of fun but also a lot of insight in both computer science and physics.

By now, we already saw one quantum algorithm: quantum FFT. This is based on yet another fundamental structure from information theory: a binary tree. Let's use quantum FFT in cryptography.

The RSA key exchange uses two long prime numbers: $p < q$ (both kept secret). Their product pq, on the other hand, is not secret at all but public. How to break the code? In other words, how to factorize pq in terms of its inner (unknown) factors, p and q? This is not easy. Why? Because this is a "salad" problem: separate your salad to its original ingredients back again!

Can you do this? This is quite difficult: much more difficult than preparing the salad in the first place... A quantum computer is needed.

This is the factoring problem: the job of the hacker. Once he/she uncovers p and q explicitly, he/she can easily read our secret message (which we don't want).

Although we don't have a quantum computer yet, we never know: maybe the hacker already has! It is quite important to understand how he/she could break our code. This will make us aware of the risk and search for a cure.

On the good old digital computer, it is practically impossible to factor pq in terms of p and q back again. To break the code, the hacker needs a quantum computer. On it, he/she should run quantum FFT, in an implicit version: no need to read the result! Even so, this is good enough to uncover p and q explicitly and break the RSA code.

Fortunately, no hacker has a quantum computer yet (or at least we hope not). Still, better get ready to a future day, when quantum computers may get common, and available to everyone. For this purpose, better understand well how to break the code. Only then can you try and close the gap and design a new (safer) code.

15.1 FFT and RSA

15.1.1 *Implicit Quantum FFT*

By now, our quantum FFT remains implicit: we can never read the result, or we'd spoil the entire state forever! How to use quantum FFT in practice? Only implicitly, within a bigger algorithm. How, and for what purpose? To break the RSA code.

15.1.2 *The Factoring Problem*

In the RSA code, in the public domain, we only place the product: pq. How to uncover its original (secret) prime factors: p and q? This is the factoring problem: it actually breaks the RSA code. On the digital computer, this is practically impossible. On the (theoretical) quantum computer, on the other hand, there is a good algorithm: Shor's factoring algorithm.

15.2 Hypercube and Its Index

15.2.1 *Quantum Computer: Hypercube*

To help solve the factoring problem, assume that we already have a quantum computer: an array of k qubits, one by one, in a row. But what is the computational domain? Well, each qubit could be either 0 or 1. Together, these (hypothetical) values make a new geometrical structure: a hypercube of 2^k nodes, indexed by

$$i = 0, 1, 2, 3, \ldots, 2^k - 1.$$

This hypercube is quite nonphysical: it is in our mind only. Still, we can imagine it and work with it.

To index the nodes in the hypercube, i must take its binary representation (configuration): a list of k binary digits (0 or 1). For this purpose, each qubit makes a little random variable, with two possible values: 0 or 1. Together, the qubits make a bigger random variable: i itself. Still, this is more complicated than a system of (independent) coins: the qubits can be dependent or at least entangled!

15.2.2 *Binary Index: A New Random Variable*

Here, i itself is a random variable too, in its own right. What is the value of i? Well, we don't know for sure: this is uncertain and can be predicted at some probability only.

What is the probability? Well, this depends on each individual qubit. If it spins up, then it has value 0. If, on the other hand, it spins down, then it has value 1. This is uncertain: each qubit makes a little random variable. Together, these k qubits might spin in such a configuration that makes i. The probability for this is $|v_i|^2$, where

v is the state: a normalized (complex) 2^k-dimensional vector (or grid function), defined for all $0 \leq i < 2^k$.

This is not as simple as a system of coins. Indeed, unlike coins, here the qubits are entangled: not quite independent. More precisely, their "little states" depend on each other. This is why v *cannot* be written as a product of k little states: it contains not only k but also $2^k - 1$ degrees of freedom. This is what gives our quantum computer its exponential power. Let's see a few examples.

15.3 Transformation: From State to State

15.3.1 *The Initial State*

Let's prepare our quantum computer in a very simple (deterministic) state: all qubits spin up (and have value 0). This way, i is deterministic too. After all, we already know its value for sure:

$$i = 0.$$

This is why the state takes a particularly simple face; only the first component is nonzero, and all the rest vanish:

$$v = (1, 0, 0, 0, 0, \ldots, 0)^t$$

(Figure 15.1). But don't worry: thanks to quantum FFT, our state will soon change dramatically and get as nondeterministic as ever — uniformly distributed. For this, v must change into a constant vector of all 1's (up to a normalization factor, which we disregard for now).

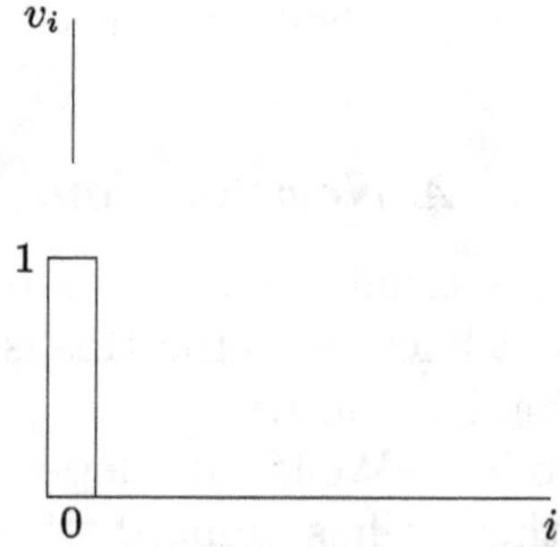

Figure 15.1. Initially, i is not random but deterministic: $i = 0$ at probability 1 (a must) and $i \neq 0$ at probability 0 (no chance at all). In other words, we know for sure that all qubits spin up, not down.

15.3.2 *The One State: Uniform Distribution*

By now, our state v is deterministic: it has only one nonzero component. But this is not what we want. On v, let's run quantum FFT. In other words, to the above v, apply the $(2^k \times 2^k)$ matrix W_{2^k}:

$$v \to (W_{2^k})\, v = (1, 1, 1, 1, \ldots, 1)^t$$

(Figure 15.2). Indeed, in W_{2^k}, there are powers of w (the 2^kth root of unity). In particular, the first column contains just $w^0 = 1$.

15.3.3 *Complete Nondeterminism*

What is this new v? It is as nondeterministic as ever. In fact, each qubit is equally likely to spin up or down. This way, i is really random: equally likely to take any configuration, with any value from 0 to $2^k - 1$.

15.4 Period Finding

15.4.1 *Modular Power*

But this is still not quite what we want. We really want a state with a "periodic zebra" pattern. To have this, recall our original task: to uncover the prime numbers $p < q$. Fortunately, we already have their product: pq. Let's pick a new natural number $1 < a < pq - 1$

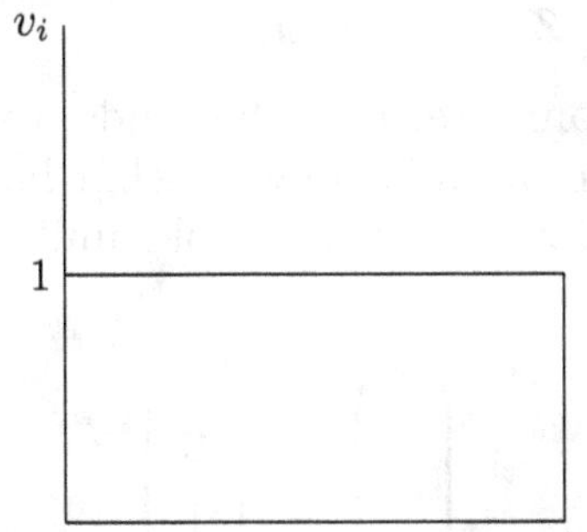

Figure 15.2. After FFT, v gets constant and uniform; all components are the same: $v_i = 1$ (up to normalization). This means complete nondeterminism: i is equally likely to take any value between 0 and $2^k - 1$. In other words, all qubits are equally likely to spin up or down.

at random. Now, is a coprime to pq (Section 12.4.1)? If not, then a must be a multiple of p or q (but not both). In this case, we're done: just calculate

$$\mathrm{GCD}\,(a, pq)\,,$$

and get both p and q explicitly, as required. So, assume that a is coprime to pq. In other words, a is *not* a multiple of p or q:

$$\mathrm{GCD}\,(a, pq) = 1.$$

Thanks to this new a, we can now calculate a new random variable: the modular power

$$a^i \bmod pq$$

(Section 12.4.3).

15.4.2 *Quantum Computing*

i is a random variable: nobody knows what it really is. Still, our quantum computer can use it to calculate this modular power and store the result. Once this is done, v changes forever: only those (hypothetical) i's that could produce this result remain valid (with $v_i = 1$), while the others are marked invalid (with $v_i = 0$). Which ones are still valid? We don't need to know. We only need to know the spacing between them (denoted by o). After all, this is what gives v its new face: the new "zebra" pattern.

15.4.3 *"Periodic Zebra" State*

Once the quantum computer uses the random variable i to calculate (and store) the above modular power, what happens to the state of i? Well, v has transformed from its old uniform pattern to a new "zebra" pattern:

$$\begin{pmatrix} 1 \\ 1 \\ 1 \\ 1 \\ 1 \\ \vdots \end{pmatrix} \rightarrow \begin{pmatrix} 1 \\ \mathbf{0} \\ 1 \\ \mathbf{0} \\ 1 \\ \vdots \end{pmatrix}.$$

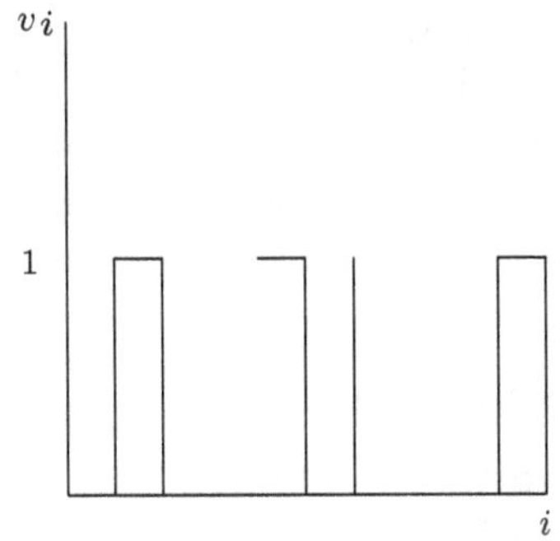

Figure 15.3. After calculating the modular power, the state takes a new "zebra" pattern: $v_i = 1$ at i's at spacing o from one to another and $v_i = 0$ in between.

(Here, **0** is a zero subvector of dimension $o - 1$: Figure 15.3.) On the left-hand side, the old v had a uniform pattern: 1's everywhere. On the right-hand side, v got a much more interesting pattern: $v_i = 1$ only at those i's for which (a^i mod pq) agrees with the stored result. At the other i's, v_i drops to zero:

$$v_i = \begin{cases} 1 & \text{if } (a^i \bmod pq) = \text{ stored result} \\ 0 & \text{otherwise.} \end{cases}$$

After all, those i's that disagree with the stored result could never take place: they have probability 0 — no chance at all.

15.4.4 *The Order of a*

Why is this a zebra pattern? To see this, recall the order of a:

$$o = o_{pq}(a).$$

This is the minimal natural number for which

$$a^o \equiv 1 \bmod pq.$$

Thus,

$$o \mid \phi = (p-1)(q-1).$$

(Indeed, in Sections 12.7.5–12.8.1, pick $m = a$.) This will help bound o but not uncover it. After all, we have no access to ϕ, which is kept secret. (Don't forget: we are now the hackers...) Thus, we have a new task: uncover o, without knowing ϕ. Why is o so important? Later on, we'll use o to solve the factoring problem. Fortunately, we already have an important clue about o: it makes the spacing in our up-to-date state.

15.4.5 *Spacing and Period*

Indeed, by now, our up-to-date v already looks like this: 1, followed by $o-1$ zeroes, followed by 1, followed by $o-1$ zeroes, and so on. More precisely, this pattern is not quite unique; it could also shift, provided that the same spacing is kept: $o-1$ zeroes between the 1's. But we don't care about this shift: it can only affect phase, not magnitude (in quantum FFT, to be applied soon).

15.5 Period Finding: Use Quantum FFT

15.5.1 *Applying Quantum FFT*

By now, the period o is still unknown. How to find it? Apply quantum FFT! In the new state v, we'll then have a new (approximate) zebra pattern:

$$v \to \begin{pmatrix} 1 \\ \mathbf{0} \\ 1 \\ \mathbf{0} \\ 1 \\ \vdots \end{pmatrix}.$$

15.5.2 *Near Zebra State*

This is a new near-zebra pattern: 1, followed by $2^k/o - 1$ zeroes, followed by 1, followed by $2^k/o - 1$ zeroes, and so on (Figure 15.4). But this is just an approximation. Why? Because, in general, $2^k/o$ is not an integer:

$$o \not| \, 2^k.$$

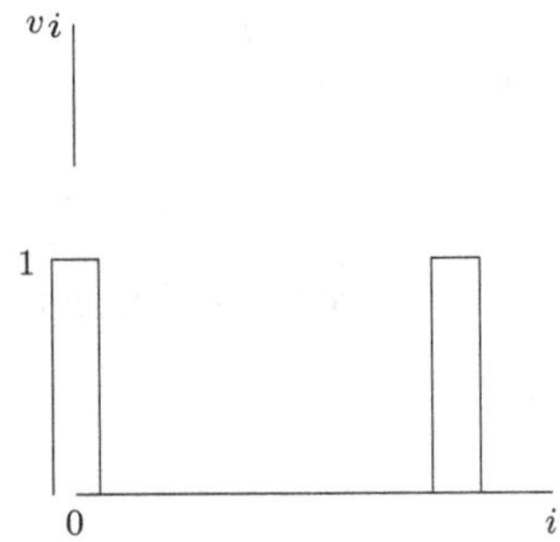

Figure 15.4. After quantum FFT, the state is near-zebra, with spacing as big as $2^k/o$.

As a matter of fact, $2^k/o$ could be written as an integer minus a (positive) fraction:

$$\frac{2^k}{o} = \left\lceil \frac{2^k}{o} \right\rceil - \alpha,$$

for some $0 < \alpha < 1$. Fortunately, we can still pick k so big that

$$2^k \gg (pq)^2 > \phi^2 \geq O^2.$$

This way, the new spacing $2^k/o$ is much bigger than the old spacing o. This way, the new random variable o is not as sensitive as the old random variable i: even if we have an error in i, this is still unlikely to inherit to o, and affect or spoil it.

15.6 Interference: Constructive or Destructive?

15.6.1 *Constructive Interference*

What produces the near-zebra pattern? To see this, consider the old v again (Figure 15.3), and apply quantum FFT to it. How should the new v look like? Look at it as a column vector (top to bottom, with increasing i), and you'll see an interference pattern: constructive, followed by destructive (below it), followed by constructive again, followed by destructive again, and so on. Indeed, consider an i that is roughly a multiple of $2^k/o$:

$$i \doteq l\frac{2^k}{o},$$

for some $0 \le l \le o$. In other words, the error is small (in magnitude):

$$e = e(i) = i - l\frac{2^k}{o} \ll \frac{2^k}{o}.$$

(This could be negative but is still small in absolute value.) After FFT, how should the ith component look like? Well, during FFT, v is highly likely to undergo constructive interference (at its ith component):

$$\begin{aligned}
2^{k/2}\left(\left(W_{2^k}\right)v\right)_i &= \sum_{j=0}^{2^k-1} w^{ij}v_j \\
&= \sum_{0\le j<2^k/o} w^{ioj} \\
&\doteq \sum_{0\le j<2^k/o} w^{l2^kj} \\
&= 1+1+1+\cdots+1 \\
&= \frac{2^k}{o} + \alpha.
\end{aligned}$$

This is indeed constructive: in this sum, all terms are positive and help increase the total sum. (Compare this to Section 2.2.1.)

15.6.2 *Interference: Constructive vs. Destructive*

And what about a more general i, which could be quite different from any multiple of $2^k/o$? Well, for such an i, during FFT, v undergoes a different kind of interference: not constructive but destructive. How to analyze both kinds at the same time? For this purpose, write e as an integer plus a roundoff error:

$$e = M + g,$$

for some

$$|g| \le \frac{1}{2}.$$

This way, we can now estimate the above sum asymptotically (for every i):

$$
\begin{aligned}
&2^{k/2}\left(\left(W_{2^k}\right)v\right)_i \\
&= \sum_{j=0}^{2^k-1} w^{ij} v_j \\
&= \sum_{0 \le j < 2^k/o} w^{ioj} \\
&= \frac{1 - w^{io(2^k/o+\alpha)}}{1 - w^{io}} \\
&= \frac{1 - w^{eo(2^k/o+\alpha)}}{1 - w^{eo}} \\
&= \frac{1 - w^{eo\alpha+e2^k}}{1 - w^{eo}} \\
&= \frac{1 - w^{eo\alpha+g2^k}}{1 - w^{eo}} \\
&\sim \begin{cases} \frac{\sqrt{(1-\cos(2\pi g))^2+\sin^2(2\pi g)}}{2\pi eo/2^k} = \frac{\sqrt{2-2\cos(2\pi g)}}{2\pi e} \cdot \frac{2^k}{o} = \frac{\sin(\pi g)}{\pi e} \cdot \frac{2^k}{o} & \text{if } eo \ll g2^k \\ O(1) & \text{if } g2^k = O(eo), \end{cases}
\end{aligned}
$$

as can be seen geometrically on the unit circle in the complex plane (Figure 15.5).

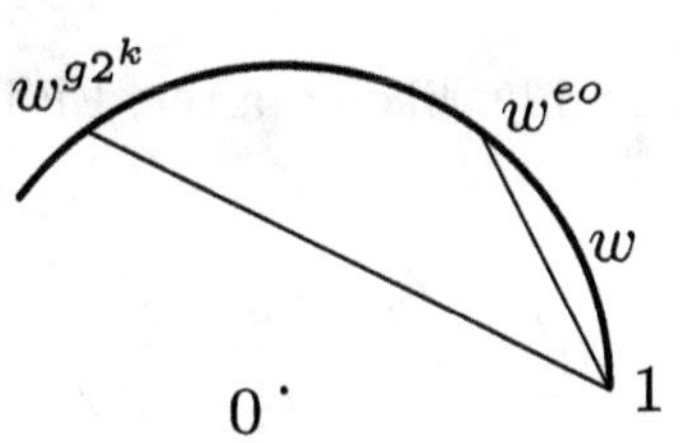

Figure 15.5. In the complex plane, w lies on the unit circle, at phase $2\pi/2^k$. Now, if $eo \ll g2^k$, then w^{g2^k} has a much bigger phase than w^{eo} and a much longer distance from 1, so $1 - w^{g2^k} \gg 1 - w^{eo}$, so the numerator is much bigger than the denominator (in magnitude). This is constructive interference, with high probability $|v_i|^2$ (after FFT).

15.6.3 *Low Error: High Probability*

Look at those i's with error as low as

$$e = O(go),$$

so

$$eo = O\left(go^2\right) \ll g2^k.$$

(Interpret these inequalities in magnitude: if e or g are negative, then they pick a minus sign.) These i's win: after FFT, they have probability as high as

$$|v_i|^2 \sim \left(\frac{\sin(\pi g)}{\pi e} \cdot \frac{2^k}{o}\right)^2$$

(up to normalization). For which i is this really high? For such an i with

$$M(i) = 0.$$

Such an i is nearly accurate: it has an error as low as

$$e(i) = g(i).$$

In other words, such an i is (nearly) a multiple of $2^k/o$:

$$i \doteq l\frac{2^k}{o}.$$

Thus, such an i is highly likely: after FFT, it has probability as high as

$$|v_i|^2 \sim \left(\frac{\sin(\pi g)}{\pi g} \cdot \frac{2^k}{o}\right)^2$$

(up to normalization). Is this result familiar? We already saw it before: in the special case of $e = g = 0$, we even saw an exact equation (Section 15.6.1).

15.6.4 *High Error: Low Probability*

And what about those i's with high error? For these, things are the other way around: the interference is not constructive but destructive. This leads to a much lower probability (up to normalization again):

$$|v_i|^2 = O(1) \quad \text{if} \quad g2^k = O(eo) \quad \text{and} \quad e \neq 0.$$

In particular, this is relevant to an i with a big error:

$$|v_i|^2 = O(1) \quad \text{if} \quad 2^k = O(eo).$$

Indeed, since $|g| \leq 1/2$, this condition is even stronger and implies the previous one.

15.7 Finding the Order of a

15.7.1 *Measuring i: Back to Determinism*

How to uncover o? By now, we have a near-zebra state. It is time to change it for the last time. For this purpose, just read (or measure) i. This makes v as simple as in the beginning: just one nonzero component:

$$v_j = \begin{cases} 1 & \text{if} \quad j = i \\ 0 & \text{if} \quad j \neq i. \end{cases}$$

This is the end of our quantum computation. We are now in determinism back again; we know i for sure, with no doubt any more: at probability $|v_i|^2 = 1$. There is no room for ambiguity any more. Indeed, after measuring i, v has just one nonzero component: v_i. All other v_j's (for any $j \neq i$), on the other hand, drop to zero. After all, these j's disagree with i, and can't be true any more.

Moreover, thanks to our asymptotic estimate, i is highly likely to be near a multiple of $2^k/o$:

$$i \doteq l\frac{2^k}{o}$$

(for some unknown integer $0 \leq l \leq o$). To uncover o, we can now use the good old digital computer back again.

15.7.2 Continued Fraction

Unfortunately, l is still unknown. As a result, o is unknown too. Although we don't care about l, we do care about o. How to uncover it? By approximating

$$\frac{i}{2^k} \doteq \frac{l}{o}$$

as a continued fraction. This is done on a digital computer: no need to use a quantum computer any more.

How to do this? Define a new sequence of natural numbers: $o_1, o_2, o_3, o_4, \ldots$, which are optimal in the following sense: o_1 is the maximal integer number that makes a lower bound:

$$o_1 \leq \frac{i}{2^k}.$$

(In our case, $o_1 = 0$.) Next, o_2 is the maximal natural number that produces an upper bound of the form

$$o_1 + \frac{1}{o_2} \geq \frac{i}{2^k}.$$

Next, o_3 is the maximal natural number that produces a new lower bound of the form

$$o_1 + \frac{1}{o_2 + \frac{1}{o_3}} \leq \frac{i}{2^k}.$$

Next, o_4 is the maximal natural number that produces a new upper bound of the form

$$o_1 + \frac{1}{o_2 + \frac{1}{o_3 + 1/o_4}} \geq \frac{i}{2^k},$$

and so on. On the left-hand side, we get a list of alternating (lower or upper) bounds. Each bound could be written as l/o, where o increases monotonically down the list, yielding better and better accuracy. The final (and best) o should be used to solve the factoring problem, as discussed in the following (provided that $o < pq$).

15.8 Breaking the RSA Code

15.8.1 *Nontrivial Square Root*

What is so good about o (the order of a)? Well, thanks to a lemma in Ref. [34], at probability 1/2 or more, o is even, and $a^{o/2}$ makes a nontrivial square root of 1 modulus pq. What does this mean? Well, to be nontrivial, this square root mustn't be ± 1 modulus pq:

$$a^{o/2} \not\equiv \pm 1 \bmod pq.$$

In other words, pq mustn't divide the following two numbers:

$$pq \not| \ a^{o/2} \pm 1.$$

And what about the product of these two numbers? Well, thanks to the original definition of o, pq must divide it:

$$pq \mid a^o - 1 = \left(a^{o/2} - 1\right)\left(a^{o/2} + 1\right).$$

15.8.2 *Factorization*

Is this situation familiar? Well, this is as in Section 12.9.4 (but with a much lower modular power):

$$p \mid a^{o/2} - 1 \quad \text{and} \quad q \mid a^{o/2} + 1$$

(or vice versa). This is what we wanted. To have p and q explicitly, just calculate

$$\text{GCD}\left(a^{o/2} - 1, pq\right).$$

Thanks to the above lemma, at probability 1/2 or more, this should work. Otherwise, pick a new a, and restart all over again.

15.9 Exercises: FFT of Any Order

15.9.1 *How to Truncate a State?*

1. Look again at quantum FFT.

2. So far, the dimension was a power of 2: 2^k. In this case, we already know how to apply FFT (the unitary matrix W_{2^k}).
3. But what if the dimension was a new natural number L that may not be a power of 2 any more? In this case, v is an L-dimensional (complex) vector. How to apply W_L to it? *Hint*: see the following.
4. Let v be a 2^k-dimensional state. How to truncate (or mask) it to have just L nonzero components? More precisely, how to modify v to keep only the L leading components, followed by $2^k - L$ zeroes? *Hint*: on your quantum computer, consider a run, during which we need to measure i, and use it. Here is an algorithm to check whether to accept the result (if $i < L$), or reject and restart (if $i \geq L$):
   - Recall that L is deterministic.
   - Look at L, in its binary representation.
   - Add leading zeroes to it (if necessary), until it has k bits as well.
   - Consider the (nondeterministic) binary representation of i, qubit by qubit.
   - Start from the most significant qubit. It spins up (value: 0) or down (value: 1). Without recording it, compare it to the corresponding bit in L.
   - If it is smaller, then accept (because $i < L$).
   - If, on the other hand, it is larger, then reject (because $i > L$).
   - If, on the other hand, it is the same, then:
     - If this is the last (least significant) qubit, then reject (because $i = L$).
     - Otherwise, move to the next (less significant) qubit, and compare it too (as before), and so on.

   This completes the algorithm.
5. Does this truncate v, as required? *Hint*: by the end of this algorithm, we still don't know what i really is. We only know whether $i < L$ or not. Only if $i < L$ do we get to continue our run. Thus, in this case, the state was indeed truncated:

$$v \to \begin{pmatrix} \mathbf{v} \\ \mathbf{0} \end{pmatrix},$$

where $\mathbf{v}$ is the top L-dimensional subvector of v and $\mathbf{0}$ is a new $(2^k - L)$-dimensional dummy subvector. Why? Because there is now no chance to have $i \geq L$ any more.

6. How to apply W_L to the top subvector $\mathbf{v}$? *Hint*: truncate v as above, and then apply W_{2^k} to it, with a few changes:
   - Instead of the 2^kth root of unity, use the Lth. Do this implicitly: use a new w, with a new phase, $2^k/L$ times as big. How?
   - For this purpose, start from a new initial frequency, $2^k/L$ times as high.
   - In the end, truncate again.
   - Finally, renormalize: multiply by $\sqrt{2^k/L}$.

   This way, the final state will be the same as $W_L\mathbf{v}$, as required.

Part VI: Feynman's Theory: An Algebraic Point of View

Quantum mechanics works gradually, step by step: It starts from statics and then moves on to dynamics too. Indeed, the original wave function tells us about static position (in 1-D). This is then extended to 3-D too, to help model angular momentum too. But what about time? Why leave it out? Better throw it in and treat it geometrically, just like any spatial dimension. This produces spacetime: a new (4-D) differentiable manifold. (To define it, use Zorn's lemma: Chapter 11 in Ref. [27].)

This geometrical point of view is the basis for quantum field theory: instead of integrating over individual points in space, integrate over complete paths in spacetime. This is the path integral: paths superpose and interfere, giving us an immediate look of the entire dynamics, on spot, with no need to wait any more!

In quantum mechanics, we already studied an individual particle and its dynamics. To see this, we looked at its wave function and how it changes in time. For this, we used our new time–energy polar coordinates (Figures 3.1 and 3.2): time is circular (or angular), and energy is radial. This plane is named "energytime." In it, how to illustrate the dynamics? Well, as time goes by, the entire energy ray rotates counterclockwise. This mirrors Schrodinger's equation, in its geometrical face.

What is the physical meaning of this? On the radial ray (issuing from the origin), there are (discrete) allowed points. On them, the discrete wave function is defined, telling us how likely the particle is to have a particular energy. Each allowed point stands for an eigenfunction of the Hamiltonian, conserved in time.

As time goes by, such a point makes a complete circle: an energy level. On it, the wave function only precesses, with no physical effect at all. This means that position–momentum remain the same, all the time (stochastically). If, on the other hand, two (or more) eigenfunctions superpose and interfere, then energy is still conserved (only stochastically now), but position–momentum have a real physical dynamics now. Why? Because the eigenfunctions precess at a different frequency, and interfere dynamically: constructively, then destructively, then constructively again, and so on.

Feynman, on the other hand, introduced dynamics in real physical spacetime. He looked at a path, leading from the initial event to the final event. The particle could follow this path or any other path connecting these two events (at some probability). In quantum field theory, we superpose (or sum, or even integrate) all possible paths.

This is the path integral. Thanks to it, we get to see how all paths interfere at the final event. We already saw a simple example: the double-slit experiment, with just two paths. (In Figure 4.1, let time flow rightward, with the particle, and space be vertical.) But, in more interesting cases, there could be infinitely many paths. Each one can be illustrated in a Feynman diagram. In it, new particles could appear, and old ones could drop (at some probability), provided that energy is still conserved along the path.

In quantum mechanics, we mostly focus on just one particle. At best, we look at two and how they entangle. But how is a new particle born? Where did it come from? To explain this, Feynman introduced quantum field theory. In it, we have new operators, to help create a new particle and annihilate it back again.

This leads to Feynman diagrams. As a matter of fact, we already saw a simple example: an electron could absorb a photon, and "jump" to a higher energy level. On the other hand, it could also spit it back, and fall to a lower energy level back again. Here, we'll see higher examples as well: two electrons could even exchange a virtual photon and repel one another. This is an electrostatic force, in its new quantum-mechanical face.

Feynman's diagrams are geometrical. Still, they also have an algebraic face: tensor product and the commutator. This will help design the transition matrix. In particle physics, this will help model weak and strong interactions alike: bosons, fermions, quarks, gluons, and more. This is a good preparation work for a more general study [23].

Chapter 16

Feynman Diagrams and The Commutator

In quantum computing, we often use a binary tree: the leaves at the bottom level model the nodes in a (virtual) binary hypercube, which makes a quantum computer. Moreover, a (virtual) binary tree is also useful to model a recursive algorithm like FFT. This helps parse the recursion, level by level, and implement it nonrecursively, and more efficiently.

A tree is a special kind of graph. Let's look at a more general graph. What is it? It is a set of nodes. On top of this, it also has a set of edges, to connect nodes. In an oriented graph, an edge has a direction: it is like an arrow, pointing from one node to another, not the other way around. Moreover, in a weighted graph, each edge also has a weight: a number, telling us how likely the edge is to get active, and let mass flow along the arrow.

Thanks to graph theory, we can now introduce Feynman diagrams [32]. These are elementary pieces of graph, containing one node (or vertex) and three edges issuing from it. Thanks to their edges, Feynman diagrams can be tied to one another and assemble bigger and bigger graphs.

What is the physical meaning of this? Well, a Feynman diagram illustrates an elementary quantum-mechanical process, in which two particles scatter from one another and emit a new particle (or absorb an old one). This could take place at some probability, calculated from the (complex) matrix of weights. This is the algebraic face: to assemble two diagrams, use the commutator.

This is quite useful in particle physics and quantum field theory. This is the algebraic face again: to unite two particles into a new system, use their tensor product, and (the commutator of) their own (complex) matrices. Thanks to their elegant rules, Feynman diagrams form a complete algebraic structure, easy to visualize and work with. This is how modern physics takes its algebraic-geometrical nature and gets easy and transparent.

16.1 System: Particle and Anti-Particle

16.1.1 *Particle and Its Anti-Particle*

In quantum mechanics, consider some particle (say, an electron). Assume that it could be in m possible states (each at some probability), where m is a fixed natural number. For an electron, for example, there are only two states: spin-up or spin-down ($m = 2$: Figure 16.1).

16.1.2 *Positron: A Missing Electron of Negative Energy*

Usually, the particle has a positive energy. Still, in theory, it could also have a negative energy. In this case, its anti-particle has positive

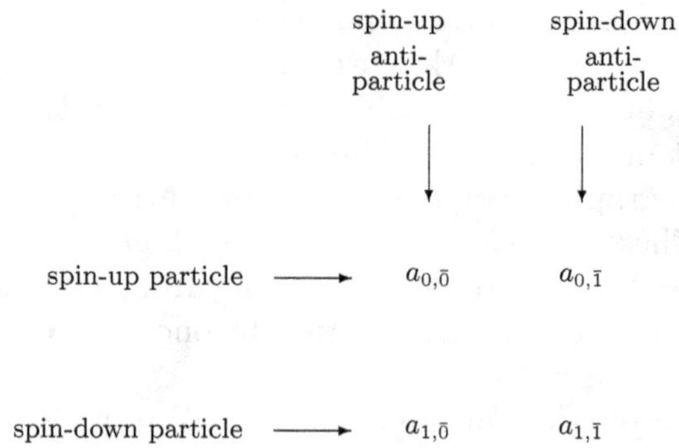

Figure 16.1. The simple case $m = 2$. The system contains a particle and an anti-particle, each could spin up or down. This is indicated by their indices: $0 \leq i, j < m = 2$. The $i\bar{j}$th possibility has complex amplitude $a_{i,\bar{j}}$. This is the joint (discrete) wave function, denoted above by $v_{i,j}$ or $v(i,j)$. Once normalized, it has probability $|a_{i,\bar{j}}|^2$ (known as the fine structure constant).

energy. For example, consider an (imaginary) electron of negative energy. But, it is missing and only leaves a hole. This hole is called positron (an anti-electron): a missing electron of negative energy. This is why it has positive energy and charge. If an electron meets a positron, then it "falls" into the hole, fills it, and releases the double energy.

The same works for a more general particle (not only an electron). What is its anti-particle? This is its mirror image:

$$\text{anti-particle} = \text{a hole, or a missing particle.}$$

This way, it has opposite energy, charge, and spin: all three pick a minus sign.

16.1.3 *Complex Amplitude*

Together, our original particle and its anti-particle form a new system, with the following $m \times m$ matrix:

$$A \equiv \left(a_{i,\bar{j}}\right)_{0 \le i,j < m}$$

(Figure 16.1). (The bar reminds us that this is an anti-particle.) What do we have in A? Each element is called a complex amplitude. Is this familiar? In previous chapters, this was called the joint (discrete) wave function (of both particles together, as one complete system) and denoted by $v_{i,j}$ or $v(i,j)$. But here, we view it as a matrix A, ready for all sorts of algebraic operations. Next, let's focus on our simple example.

16.1.4 *Example: Electron–Positron*

To make things a little more concrete, let's focus on our (theoretical) example: a pair — electron–positron. Here, $0 \le i,\ j < m = 2$. The electron could spin either up ($i = 0$) or down ($i = 1$). Likewise, the positron could spin either up ($j = 0$) or down ($j = 1$). So, the system has four possible states: either up-up or up-down or down-up or down-down.

On its own, each of these four is deterministic, with no ambiguity or uncertainty at all. Still, in A, they are superposed into a more interesting (nondeterministic) state: each one is possible, at its own complex amplitude and probability. This is our new 2×2 complex matrix:

$$A \equiv \begin{pmatrix} a_{0,\bar{0}} & a_{0,\bar{1}} \\ a_{1,\bar{0}} & a_{1,\bar{1}} \end{pmatrix}$$

(Figure 16.1). In A, each element is a complex amplitude. For example, up-down has complex amplitude $a_{0,\bar{1}}$. This tells us the probability to see up-down: $|a_{0,\bar{1}}|^2$. But these four probabilities must sum to 1, so better assume that A was already normalized:

$$\sum_{i,j=0}^{m-1} |a_{i,\bar{j}}|^2 = 1.$$

16.1.5 *Virtual Gauge Boson*

The same matrix will also serve as a discrete wave function for a boson. To motivate this, imagine a theoretical (nonphysical) situation: the electron "sits" on top of the positron. Why don't they cancel each other? Well, maybe they do! When? We don't know! In fact, thanks to the time–energy uncertainty principle, time is too short to distinguish between existence and nonexistence. The result is a virtual particle that fluctuates ever so fast, to an electron–positron pair and back, and so on. This is a gauge boson: a virtual photon.

In terms of spin, what is the state of our photon? Well, this depends on our original electron–positron. Still, could it be deterministic (say, up-up)? No! It must be nondeterministic: a superposition of up-up *minus* down-down. This means uncertainty: either up-up or down-down (at the same probability). In this virtual photon, there are two processes, illustrated by arrows. At the top arrow, the electron ed fluctuates from up to down (and back):

$$\begin{matrix}\text{electron} & - \\ \text{positron} & -\end{matrix} \quad \begin{pmatrix} 0 \\ \bar{0} \end{pmatrix} \rightleftarrows \begin{pmatrix} 1 \\ \bar{1} \end{pmatrix} \quad \begin{matrix}- & \text{electron} \\ - & \text{positron.}\end{matrix}$$

At the bottom arrow, on the other hand, the positron fluctuates from down to up (and back). But this runs the other way around: back-in-time. Later on, we'll see why.

To get balanced, the top arrow must emit a new $0\bar{1}$-photon. The bottom arrow, on the other hand, absorbs the $1\bar{0}$-anti-photon (back-in-time). This preserves energy: after all, both photon and anti-photon have the same energy. Thanks to the minus sign, time can go on (as in Schrodinger's equation). Still, quantum-mechanically, time is only uncertain: "back-in-time" is only theoretical.

Thanks to the frequent fluctuations, we have uncertainty: at a given time, we can't tell whether it is up-up or down-down. After all, both are equally likely. Still, this explanation is a bit too geometrical. How to see this algebraically too?

16.2 Matrix of Complex Amplitudes

16.2.1 *Traceless Matrix*

So far, we modeled a superposition: up-up minus down-down. Why does down-down pick a minus sign? Physically, this lets time go on (as in Schrodinger's equation) and carry out the virtual process. Algebraically, on the other hand, thanks to this minus sign, A takes the form

$$A = \frac{\sqrt{-1}}{\sqrt{2}} \begin{pmatrix} 1 & 0 \\ 0 & -1 \end{pmatrix}.$$

Here, the coefficient has two parts. The numerator $\sqrt{-1}$ makes A anti-Hermitian. Later on, we'll see why. The denominator $\sqrt{2}$, on the other hand, normalizes A (as we did with v in previous chapters). Now, this is a diagonal matrix. Look at its main-diagonal elements. Thanks to the minus sign, they sum to zero. This means that A is traceless (or trace-free): it has zero trace. This is thanks to the minus sign.

Since A is anti-Hermitian, $\exp(A)$ is unitary (norm-preserving). This was the key to dynamics and evolution: the Liouville operator (Sections 11.6.2–11.6.4). Better yet, since A is traceless, $\exp(A)$ has determinant 1. Indeed, thanks to the Jordan form, the determinant is the product of eigenvalues (with their multiplicity). Thus, since $\exp(A)$ commutes with A,

$$\det\left(\exp(A)\right) = \exp\left(\text{trace}(A)\right) = \exp(0) = 1.$$

Thus, thanks to exponentiation, the original Lie algebra of (traceless) anti-Hermitian matrices generates a new Lie group of symmetries: the (special) unitary group.

16.2.2 *Anti-Hermitian Matrix*

Physically, why should A be traceless? To exclude a deterministic state (like up-up on its own). But this is not enough: what about up-down on its own? How to exclude it too? Require that A is also anti-Hermitian. This will force superpose: up-down $\pm$ down-up. More precisely,

$$A = \frac{1}{\sqrt{2}}\begin{pmatrix} 0 & 1 \\ -1 & 0 \end{pmatrix} \quad \text{or} \quad \frac{\sqrt{-1}}{\sqrt{2}}\begin{pmatrix} 0 & 1 \\ 1 & 0 \end{pmatrix}.$$

Are these familiar? These are the good old Pauli matrices: σ_z, σ_y, and σ_x (times $\sqrt{-1}/\sqrt{2}$, to get anti-Hermitian and normalized).

16.3 Anti-Particle and Its Anti-Perspective

16.3.1 *Anti-Particle: Opposite Charge and Spin*

Still, this is just an example. In general, why should A be anti-Hermitian? This is the algebraic face of time evolution. Indeed, thanks to Schrodinger's equation, we have the time factor

$$\exp\left(\sqrt{-1}\omega t\right) = -\left(-\exp\left(-\sqrt{-1}\omega(-t)\right)\right).$$

Physically, $-t$ means back-in-time. (This way, you can now substitute t for $-t$ back again. To pay for this, frequency already picked a minus sign.) Thanks to this, charge picks a minus sign too. After all, the positron is a missing electron: a missing negative charge or an existing positive charge. In fact, charge comes from symmetry under a global phaseshift. In both classical and quantum physics, this symmetry takes place while cycling in the complex plane, which gives the original wave its familiar sine/cosine look. (Don't confuse this with spin, which cycles in 3-D, perpendicular to the direction of motion.)

In the above formula, the two inner minus signs lead to the complex conjugate. Geometrically, this means clockwise (rather than counterclockwise) cycling in the complex plane (Figure 3.1). Besides, there are also two outer minus signs: $-\exp()$ tells us that an anti-particle is a missing particle (that could superpose with the original particle and vanish). To pay for this, we also need the outermost minus sign beforehand. This interchanges existence with nonexistence: to appear means to disappear, and vice versa. This gives the anti-particle its new (subjective) coordinates in spacetime: $-x$-$-t$ (rather than x-t). Indeed, while the original particle leaves x_0 to go to x_1, the anti-particle goes the other way around: it arrives at x_0 even before leaving x_1. Moreover, in 3-D, all three coordinates pick a minus sign, and become left-handed. This is why spin picks a minus sign too. (To see this in the original x-t coordinates, rewind the movie, to see spin go the other way around.)

16.3.2 *Why Anti-Hermitian?*

What is the algebraic effect of this? Consider an $i\bar{j}$-boson, with complex amplitude $a_{i,\bar{j}}$. Now, interchange i and j:

$$\{i \leftrightarrow j\} \Leftrightarrow \{\text{electron} \ \leftrightarrow \ \text{positron}\} \Leftrightarrow \{\text{boson} \ \leftrightarrow \ \text{anti-boson}\}.$$

This should lead to its anti-boson: a $j\bar{i}$-anti-boson. What is its complex amplitude? Start from $a_{i,\bar{j}}$, and modify it. First, it must pick a complex conjugate on top. Physically, this means negative energy and back-in-time. Next, it must also pick an outer minus sign beforehand (to pay for the fact that an anti-boson is a missing boson). Physically, this means going the other way around (in space too): absorbed rather than emitted. Thus, this is back-in-spacetime: in both space and time alike. But actually, in spacetime, each point is an event: an arrow pointing at it means that something appeared (was created) at this event. In particular, an arrow pointing downward tells us the perspective of an anti-particle. From our original perspective, on the other hand, time flows upward, so we should go against the arrow. Anyway, the algebraic result is

$$a_{j\bar{i}} = -\bar{a}_{i\bar{j}},$$

as required.

16.3.3 *Dual Perspective: Virtual Process*

This is why A must be anti-Hermitian. This way, emitting a boson is like absorbing an anti-boson, back-in-time. This is the anti-perspective. You can side with the anti-boson and take its own point

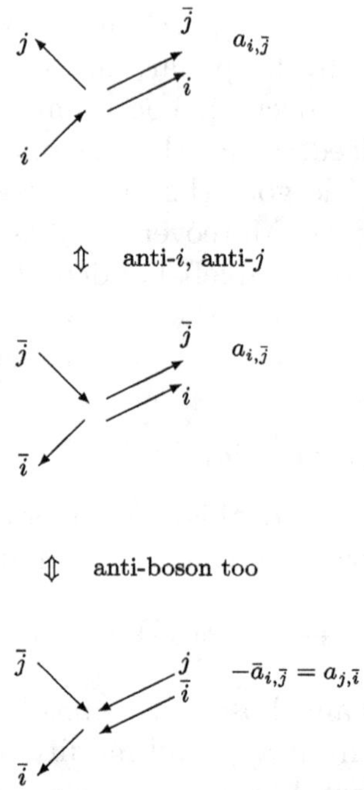

Figure 16.2. A Feynman diagram, to model scattering in spacetime: an $i \to j$-transition is like an $\bar{i} \leftarrow \bar{j}$-anti-transition (back in time). By superposing both, we get the virtual process in Section 16.1.5.

of view. The matrix A wouldn't care: it cares for spin-transition only. By superposing both perspectives, we get our virtual process back again (Section 16.1.5). Later on, this will also lead to the commutator (Figure 16.2).

16.4 Transition: Feynman Diagram in Spacetime

16.4.1 *Spin Conservation in Scattering*

Assume that an i-particle hits an object, bounces off, and changes. This is scattering. In Chapter 17, we'll see how this preserves momentum and energy. Does it preserve spin too? It sure does.

How to see this algebraically geometrically? Take an i-particle, and transition it into a new j-particle:

$$i \to j$$

(Figure 16.2). To preserve spin, an $i\bar{j}$-gauge boson is also emitted:

$$i = j + i\bar{j}.$$

Indeed, in terms of spin, j and $\bar{j}$ cancel out. In terms of energy, on the other hand, they add up and release the extra energy coming from the transition.

16.4.2 *Absorbed Anti-Boson: Back-in-Time*

In terms of anti-matter, on the other hand, this goes the other way around:

$$\bar{j} \rightarrow \bar{i}$$

(Figure 16.2). For this, a $j\bar{i}$-anti-boson is absorbed (back-in-time). As before, this preserves spin:

$$\bar{j} + j\bar{i} = \bar{i}.$$

This is the anti-perspective, which is as legitimate. In terms of spin, it has complex amplitude:

$$a_{j\bar{i}} = -\bar{a}_{i\bar{j}}.$$

How does the transition look like geometrically? Well, to go back-in-time, it should actually point leftward, not rightward:

$$\bar{i} \leftarrow \bar{j}.$$

Better yet, in a Feynman diagram (in spacetime), the arrow should point a little downward: back-in-time.

16.5 Transition: An Algebraic Picture

16.5.1 *Transition Matrix*

Look at our original transition again. How to draw it better, not only geometrically but also algebraically? Throw the complex amplitude in:

$$i \rightarrow ja_{i\bar{j}}.$$

Here, what is i? It is a (deterministic) state of the particle: an m-dimensional (standard unit) row vector, with 1 at its ith component, and 0 elsewhere:

$$i \sim (0, 0, \ldots, 0, \overset{i}{1}, 0, 0, \ldots, 0).$$

Likewise, j is a (standard unit) row vector too. This way, the row vector i transitions into the new row vector j (times $a_{i,\bar{j}}$), as required. Still, why transition to j only? Let's transition to all possible j's and superpose over j. This yields the ith row in A:

$$i \to \sum_{j=0}^{m-1} j a_{i\bar{j}} = iA.$$

Next, extend this (linearly) once again, and superpose over i as well:

$$v \to vA,$$

where v is a more general (m-dimensional) row vector: a new (nondeterministic) state of the particle.

16.5.2 *Invariant Anti-Perspective*

Next, let's work the other way around (from the anti-perspective):

$$\bar{j} \to -\bar{a}_{i\bar{j}}\bar{i}$$

(Figure 16.2). Here, $\bar{j}$ is a (deterministic) state of the anti-particle: an m-dimensional (standard unit) column vector, with 1 at the jth component, and 0 elsewhere:

$$\bar{j} \sim (0, 0, \ldots, 0, \overset{j}{1}, 0, 0, \ldots, 0)^t.$$

But then again, why transition to $\bar{i}$ only? Let's transition to all possible $\bar{i}$'s, and superpose over $\bar{i}$. This yields the $\bar{j}$th column in $-\bar{A}$:

$$\bar{j} \to -\sum_{i=0}^{m-1} \bar{a}_{i\bar{j}}\bar{i} = -\bar{A}\bar{j}.$$

Better yet, extend this (linearly) once again, and superpose over $\bar{j}$ as well:

$$v \to -\bar{A}v,$$

where v is now an (m-dimensional) column vector: a new (nondeterministic) state of the anti-particle (going back-in-spacetime). To this, apply the transpose, and interpret v as a row vector (for the anti-particle):

$$v \to -v\bar{A}^t = -vA^* = vA$$

(thanks to the anti-Hermitian property). This looks the same as before: invariant in terms of perspective. Thus, A is perspective-blind: it cares for spin-transition only, regardless of perspective.

16.5.3 *Anti-Hermitization*

To model real physics, A must be anti-Hermitian. Still, let's consider even a more general A (not necessarily anti-Hermitian or traceless). This will make our $j\bar{i}$-boson more independent: not necessarily an anti-$i\bar{j}$-boson any more. This way, we can even imagine a (nonphysical) deterministic boson and study it.

Still, if A is *not* anti-Hermitian, how to anti-Hermitize it? Subtract its Hermitian adjoint. This will produce a new anti-Hermitian matrix:

$$A - A^*,$$

as required. Physically, this means superposing both perspectives: the original perspective, plus the anti-perspective. Later on, we'll see that this works for fermions. For bosons, on the other hand, do something else: add the Hermitian adjoint, and multiply by $\sqrt{-1}$. This will produce

$$\sqrt{-1}\,(A + A^*).$$

This is anti-Hermitian but not necessarily traceless any more. (To make it traceless too, subtract its trace/m from the main diagonal.) Later on, we'll use this for bosons.

16.5.4 *Transition and Its Probability*

Look at Figure 16.2 again. At the top, we draw the original transition: $i \to j$. This could happen at complex amplitude $a_{i,\bar{j}}$, and probability

$$|a_{i,\bar{j}}|^2 \Big/ \sum_{k,l=0}^{m-1} |a_{k,\bar{l}}|^2.$$

At this probability, you'd get to see an $i\bar{j}$-boson. This way, A serves as not only a matrix but also a (discrete) wave function (denoted by $v_{i,j}$ in previous chapters).

16.6 Two Bosons Interact

16.6.1 *Transition of a Boson*

Look at our original transition again: $i \to j$ (Figure 16.2). Parallel to it, draw yet another (trivial) "transition:" $\bar{k} \to \bar{k}$. This makes a new transition (of a complete boson): an incoming $i\bar{k}$-boson (of matrix A) transitions into an outgoing $j\bar{k}$-boson (of matrix B). This only changes i to j, leaving $\bar{k}$ the same. But don't worry: later on, in the next step, you'll get to work from the other side and change $\bar{k}$ as well, if you like.

Look at the outgoing boson again: $j\bar{k}$ (Figure 16.3). Now, reverse its direction in spacetime. For this, look at its anti-boson: an (incoming) $k\bar{j}$-anti-boson, coming from the upper-left, and absorbed (rather than emitted), back-in-time. This is equivalent and keeps the same matrix: B. On the upper-right, the outgoing boson will then have a

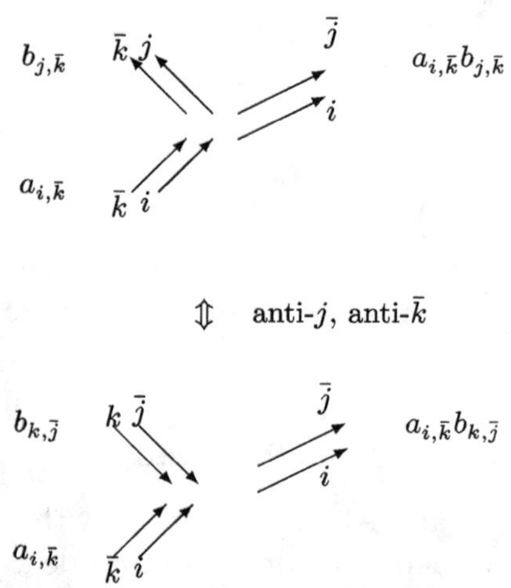

Figure 16.3. An incoming $i\bar{k}$-boson (of matrix A) transitions into an outgoing $j\bar{k}$-boson (of matrix B), emitting a new $i\bar{j}$-gauge boson. Moreover, the outgoing $j\bar{k}$-boson is equivalent to an incoming $k\bar{j}$-anti-boson, back-in-time (same matrix: B).

new complex amplitude (a new product):

$$a_{i,\bar{k}} b_{k,\bar{j}}.$$

Better yet, do the same for all possible k's, and superpose over k. On the upper-right, the outgoing boson will then take a new complex amplitude (sum of products):

$$\sum_{k=0}^{m-1} a_{i,\bar{k}} b_{k,\bar{j}} = (AB)_{i,\bar{j}}.$$

This will be quite useful in the following.

16.6.2 *Virtual Gauge Boson: Commutator*

From here, how to proceed? Well, there are two options. In the first approach, do the same at the bottom too: reverse the arrows, to point downward (rather than upward). This still keeps the same matrix: A. Now, look at the complete anti-transition: from the upper-left to the center (matrix: B), and then to the very bottom (matrix: A). As we saw above, its matrix is the product: BA. Give it a minus sign, and superpose it with the original transition:

$$AB - BA = [A, B].$$

This is the commutator: the matrix of the virtual gauge boson (emitted on the upper-right). Algebraically, it anti-Hermitizes AB (Section 16.5.3). Physically, on the other hand, it models an uncertain virtual process: each element in it would lead to some virtual process (as in Section 16.1.5), at some probability. Still, this is a bit too virtual and theoretical. Is there a more physical kind of emission?

16.6.3 *Two Bosons Emit a New Boson*

Let's go back to Figure 16.3 and take another approach. By now, we have two incoming bosons. At the bottom, the incoming boson is as in the beginning (matrix: A). The other boson, on the other hand, comes from the upper-left, back-in-time (matrix: B). Let's reverse its time. This is no longer equivalent: its matrix picks a complex conjugate on top and becomes $\bar{B}$ (Figure 16.4). And what about the

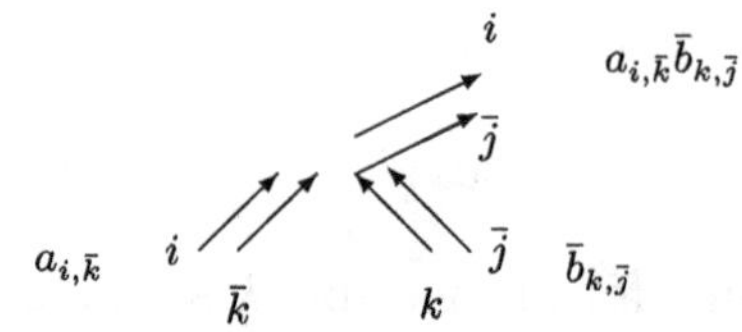

Figure 16.4. Incoming $i\bar{k}$- and $k\bar{j}$-bosons (of matrices A and $\bar{B}$) interact and emit a new $i\bar{j}$-boson. Thanks to superposition (over all possible k's), it will have a new matrix: AB (anti-Hermitized).

outgoing boson (on the upper-right)? What is its complex amplitude? It is the new product

$$a_{i,\bar{k}}\bar{b}_{k,\bar{j}}.$$

Better yet, superpose over all possible k's and sum up:

$$\sum_{k=0}^{m-1} a_{i,\bar{k}}\bar{b}_{k,\bar{j}} = \left(A\bar{B}\right)_{i,\bar{j}}.$$

For simplicity, substitute B for $\bar{B}$. This way, two incoming bosons (of matrices A and B) interact and emit an outgoing boson of a new matrix: AB? Not quite.

16.6.4 *The Commutator*

Still, there is a problem. Both A and B are legitimate: anti-Hermitian and traceless. But their product AB is not. How to fix it? Anti-Hermitize (Section 16.5.3): subtract the Hermitian adjoint:

$$AB-(AB)^* = AB-B^*A^* = AB-(-B)(-A) = AB-BA = [A,B].$$

This is the commutator: anti-Hermitian and traceless, as required. Still, there is a problem. What if

$$A = B?$$

In this case,

$$[A,A] = A^2 - A^2 = (0)$$

(the zero matrix). This means that nothing is emitted. In Section 16.6.2, this makes sense: a trivial transition like $A \to A$ should indeed emit nothing. But here, two physical (identical) bosons went in and nothing came out! Where did the extra energy (and spin) go?

16.6.5 *Fermions: Pauli's Exclusion Principle*

Thus, the commutator is good enough for fermions but not for bosons. Indeed, fermions have two interesting properties:

- On one hand, they are like bosons: indistinguishable. Thus, it makes sense to interchange them and superpose BA too.
- On the other hand, they are unlike bosons: BA picks a minus sign. This is the algebraic face of Pauli's exclusion principle; fermions can never be identical:

$$A \neq B.$$

This avoids the above paradox. Thus, for fermions, the commutator $[A, B]$ is indeed legitimate (even for two incoming fermions).

In Pauli's exclusion principle, "identical" means not only the same spin but also the same position. Only this is forbidden (for fermions). Why? Because the extra incoming energy (and spin) have nowhere to go (or scatter). But even fermions could live side-by-side in peace (not on top of one another), even with the same spin (modeled by the same matrix).

16.6.6 *Indistinguishable Bosons*

But what about our bosons?

- On one hand, they are like fermions: indistinguishable.
- On the other hand, they are not: they obey no exclusion principle at all.

Thus, once interchanged, BA picks no minus sign any more. How to anti-Hermitize? Use two steps (Section 16.5.3):

- First, Hermitize (add the Hermitian adjoint):

$$AB + (AB)^* = AB + B^*A^* = AB + (-B)(-A) = AB + BA.$$

- Then, multiply by $\sqrt{-1}$:

$$\sqrt{-1}(AB + BA)$$

(Figure 16.5). This is indeed anti-Hermitian, as required. Physically, this makes sense for bosons: indistinguishable but with no

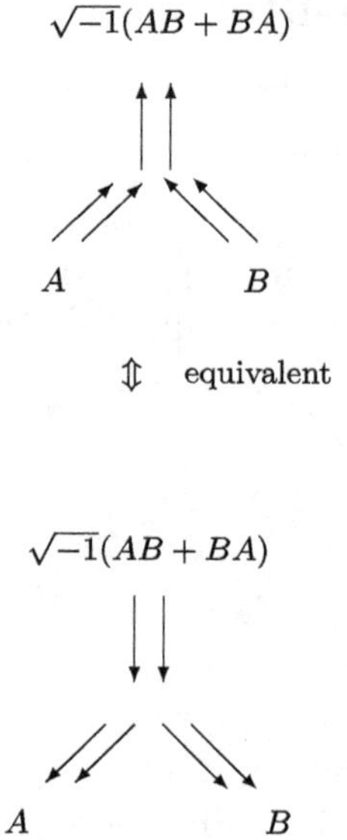

Figure 16.5. Two incoming bosons (of matrices A and B) unite and emit a new outgoing boson. Of what matrix? Well, the original bosons are indistinguishable but obey no exclusion principle. Thus, they can interchange and superpose (with no minus sign): $AB + BA$. To anti-Hermitize, multiply by $\sqrt{-1}$. Finally, this is also equivalent to the anti-perspective [same matrix: $\sqrt{-1}(AB + BA)$].

exclusion principle at all. In fact, many identical bosons could "sit" (or superpose) on top of one another, in a tall "tower." All have the same position and sign: no minus sign is needed any more.

16.6.7 *A Boson Splits into Two New Bosons*

In Figure 16.5, we have two equivalent perspectives. At the top, the process is as above: two incoming bosons unite and emit an outgoing boson. At the bottom, on the other hand, we have the anti-perspective, which is as legitimate.

Next, reverse time. This reverse-engineers the process: the entire picture turns upside-down (Figure 16.6), and the matrices pick a complex conjugate on top. Physically, the incoming boson splits into two outgoing bosons. Thanks to our matrices, this is quite simple and algebraic.

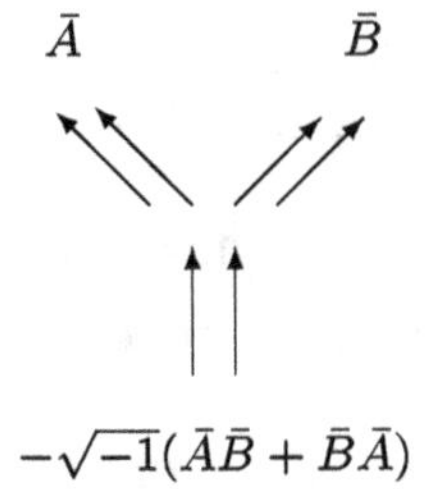

Figure 16.6. Now, reverse time. This reverse-engineers the above process: the entire picture turns upside-down, and the matrices pick a complex conjugate on top.

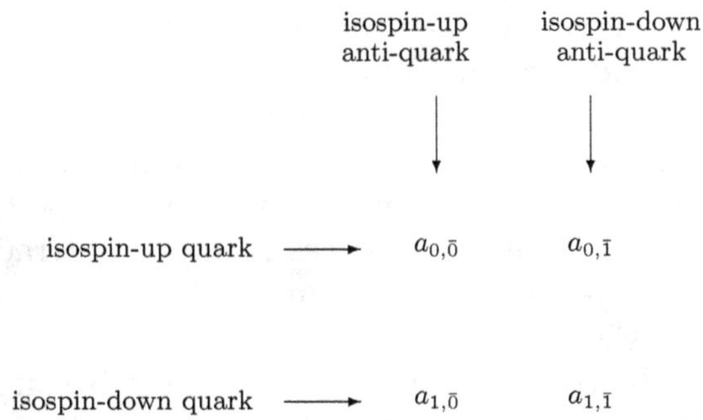

Figure 16.7. Weak interaction: a system of quark and anti-quark. Instead of spin, look now at an isomorphic property: isospin. Like spin, it could be up or down, as indicated by its index: $0 \le i, j < m = 2$. The $i\bar{j}$-boson (W-boson or Z-boson) helps transition $i \to j$ (or $\bar{j} \to \bar{i}$). It has complex amplitude $a_{i\bar{j}}$ and could be emitted at probability $|a_{i\bar{j}}|^2$ (normalized), called fine structure constant.

16.7 Exercises: Weak and Strong Interaction

16.7.1 *Weak Interaction: Quarks and W-Bosons*

1. In the atom, in the nucleus, there are protons and neutrons (which are both fermions).
2. Each proton (or neutron) contains three quarks (which are fermions in their own right).
3. Each quark has a property called isospin (which mirrors spin): either up (0) or down (1).
4. Together, a quark and an anti-quark form a new kind of boson: a W-boson. It has an isospin too: up-up ($0\bar{0}$) or up-down ($0\bar{1}$) or down-up ($1\bar{0}$) or down-down ($1\bar{1}$) (Figure 16.7).

5. Physically, what does an $i\bar{j}$-W-boson do? *Hint*: in our original quark-anti-quark system, it helps transition $i \to j$ (or $\bar{j} \to \bar{i}$, back-in-time).
6. What is its complex amplitude? *Hint*: $a_{i,\bar{j}}$.
7. At what probability could it be emitted? *Hint*: $|a_{i,\bar{j}}|^2$ (normalized).
8. This is its fine structure constant.
9. In the nucleus, what keeps the protons and neutrons together? *Hint*: this is the weak interaction: a virtual process — the proton turns into a neutron and back.
10. How? *Hint*: an up-quark from the proton turns into a down-quark, emitting a new $0\bar{1}$-W-boson (a virtual gauge boson, as in Figure 16.2 and Section 16.1.5).
11. As a result, what happens to our proton? *Hint*: this is not only a proton but also a neutron: a (nondeterministic) superposition, equally likely to be a proton or a neutron.
12. Could two W-bosons (of matrices A and B) attract each other as well? *Hint*: yes, by emitting a new W-gauge boson (of matrix $[A, B]$, as in Figure 16.3).
13. How? *Hint*: by transitioning an i-quark from boson A into a new j-quark in boson B (or a $\bar{j}$-anti-quark from boson B into an $\bar{i}$-anti-quark in boson A, back-in-time). This way, $i\bar{i}$ could even fluctuate to $j\bar{j}$ and back. (For instance, $0\bar{0}$ to $1\bar{1}$, as in Section 16.1.5.)
14. What is the probability for this? *Hint*: $|[A, B]_{i,\bar{j}}|^2$ (assuming that both A and B are normalized to have sum-of-squares 1).

16.7.2 *Weak Structure Constants*

1. Recall the Pauli matrices (Section 10.4.1):

$$\sigma_x \equiv \begin{pmatrix} 0 & 1 \\ 1 & 0 \end{pmatrix},$$

$$\sigma_y \equiv \sqrt{-1}\begin{pmatrix} 0 & -1 \\ 1 & 0 \end{pmatrix},$$

$$\sigma_z \equiv \begin{pmatrix} 1 & 0 \\ 0 & -1 \end{pmatrix}.$$

2. Are they traceless?
3. Are they Hermitian?
4. Do they anti-commute?
5. Are they cyclic? *Hint*:

$$\sigma_x \sigma_y = \sqrt{-1}\sigma_z,$$
$$\sigma_y \sigma_z = \sqrt{-1}\sigma_x,$$
$$\sigma_z \sigma_x = \sqrt{-1}\sigma_y.$$

6. How to anti-Hermitize them? *Hint*: multiply by $\sqrt{-1}$, and define three new matrices:

$$C_1 \equiv \sqrt{-1}\sigma_x = \sqrt{-1}\begin{pmatrix} 0 & 1 \\ 1 & 0 \end{pmatrix},$$

$$C_2 \equiv \sqrt{-1}\sigma_y = \begin{pmatrix} 0 & 1 \\ -1 & 0 \end{pmatrix},$$

$$C_3 \equiv \sqrt{-1}\sigma_z = \sqrt{-1}\begin{pmatrix} 1 & 0 \\ 0 & -1 \end{pmatrix}.$$

7. Are they still traceless?
8. Are they anti-Hermitian?
9. Do they still anti-commute? *Hint*: they do. After all, they are just a scalar multiple of the Pauli matrices, which do.
10. Show that they are anti-cyclic:

$$C_3 C_2 = C_1,$$
$$C_2 C_1 = C_3,$$
$$C_1 C_3 = C_2.$$

Hint: they are just $\sqrt{-1}$ times the Pauli matrices.

11. Conclude that they mirror (or represent) vector product in 3-D.
12. Pick A to be either C_1 or C_2 or C_3.
13. Likewise, pick B to be either C_1 or C_2 or C_3. (Later on, this will be extended yet more.)
14. Span their commutator as

$$[A, B] = c_1 C_1 + c_2 C_2 + c_3 C_3,$$

for some real coefficients c_1, c_2, and c_3.

15. c_1, c_2, and c_3 are called structure constants. They depend on A, B, and the original definition of C_1, C_2, and C_3:

$$c_1 \equiv c_1\left(A, B, C_1, C_2, C_3\right),$$
$$c_2 \equiv c_2\left(A, B, C_1, C_2, C_3\right),$$
$$c_3 \equiv c_3\left(A, B, C_1, C_2, C_3\right).$$

16. How could W-boson A attract W-boson B? *Hint*: by transitioning an i-quark from A into a new j-quark in B (or a $\bar{j}$-anti-quark from B into an $\bar{i}$-anti-quark in A, back-in-time).
17. What is the probability for this? *Hint*: $|[A, B]_{i,\bar{j}}|^2$ (assuming that both A and B are normalized to have sum-of-squares 1).
18. What does this tell you about c_1 and c_2? *Hint*: at least one of them is nonzero. This way, $[A, B]$ has a nonzero off-diagonal element, as required.
19. To make this transition come true, what should be emitted? *Hint*: a mixed W-gauge boson: an $i\bar{j}$-W-boson, with $i \neq j$. For instance, a $0\bar{1}$-W-boson could convert an isospin-up quark in A into an isospin-down quark in B (as in Section 16.1.5).
20. Explain this transition algebraically, and write its probability in terms of c_1, c_2, and c_3. *Hint*: the transition is uncertain. Thanks to Figure 16.3, it has its complex amplitude in AB. Once superposed with $-BA$ (the anti-perspective), this will produce $[A, B]$ (Section 16.6.2). To it, only C_1 and C_2 contribute an off-diagonal element (to help emit a mixed W-gauge boson), not C_3. Thus, the probability for this is

$$|c_1|^2 + |c_2|^2$$

(assuming that both A and B are normalized to have sum-of-squares 1).
21. Look at c_3. What does it tell you? *Hint*: c_3 is the complex amplitude of a W-boson that fluctuates from $0\bar{0}$ to $1\bar{1}$ and back:

$$W\text{-boson} \; - \; \begin{pmatrix} 0 \\ \bar{0} \end{pmatrix} \rightleftarrows \begin{pmatrix} 1 \\ \bar{1} \end{pmatrix} \; - \; W\text{-boson}$$

(Section 16.1.5). But c_3 is too selfish: it doesn't help the original W-bosons (of matrices A and B) attract one another. Why? Because it lacks an off-diagonal term, required to emit a

$0\bar{1}$-W-boson (as in its own top arrow) or absorb a $1\bar{0}$-W-anti-boson (as in its own bottom arrow).

22. So, its own arrows remain inner and can't help bosons A and B attract one another. Still, do these arrows preserve spin? *Hint*: they sure do. At the top arrow,

$$0 = 1 + 0\bar{1}.$$

And, at the bottom arrow,

$$\bar{1} + 1\bar{0} = \bar{0}$$

(back-in-time).

16.7.3 *Quantum Chromodynamics: Quarks and Gluons*

1. So far, we used $m = 2$. In this context, a quark has only two options: either isospin up or down.
2. Next, let's move on to a more complicated system, in which the quark has three options to be colored: either red (0) or green (1) or blue (2), at some probability each (Figure 16.8).

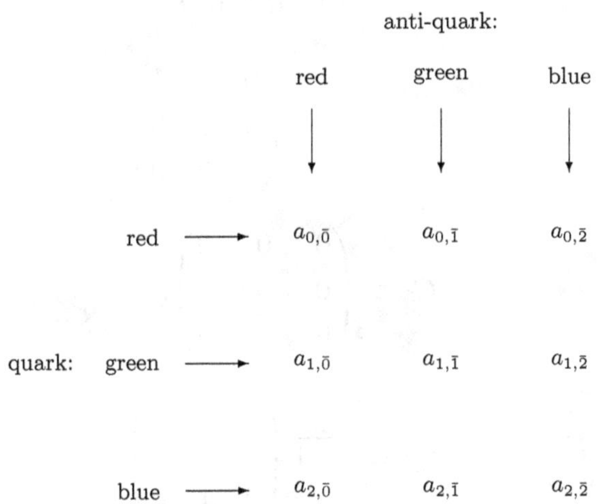

Figure 16.8. Strong interaction: a new system of quark and anti-quark, with their nondeterministic (hypothetical) color, indexed by $0 \le i, j < m = 3$. To convert the quark from color i to j, there is a need to emit an $i\bar{j}$-gluon. This has complex amplitude $a_{i\bar{j}}$, and could happen at probability $|a_{i\bar{j}}|^2$ (normalized), called fine structure constant.

3. In a proton (or a neutron), there are three quarks. What keeps them together? *Hint*: this is the strong interaction. For instance, a red quark changes to green, emitting a new $0\bar{1}$-gluon (as in Figure 16.2). This attracts the quarks to each other.
4. Could two gluons attract each other as well? *Hint*: yes (as in Figure 16.3).
5. Instead of the original Pauli matrices, what do we have here? *Hint*: define eight new (independent) 3×3 matrices:

$$C_1 \equiv \begin{pmatrix} 0 & 1 & 0 \\ -1 & 0 & 0 \\ 0 & 0 & 0 \end{pmatrix},$$

$$C_2 \equiv \sqrt{-1} \begin{pmatrix} 0 & 1 & 0 \\ 1 & 0 & 0 \\ 0 & 0 & 0 \end{pmatrix},$$

$$C_3 \equiv \begin{pmatrix} 0 & 0 & -1 \\ 0 & 0 & 0 \\ 1 & 0 & 0 \end{pmatrix},$$

$$C_4 \equiv \sqrt{-1} \begin{pmatrix} 0 & 0 & 1 \\ 0 & 0 & 0 \\ 1 & 0 & 0 \end{pmatrix},$$

$$C_5 \equiv \begin{pmatrix} 0 & 0 & 0 \\ 0 & 0 & 1 \\ 0 & -1 & 0 \end{pmatrix},$$

$$C_6 \equiv \sqrt{-1} \begin{pmatrix} 0 & 0 & 0 \\ 0 & 0 & 1 \\ 0 & 1 & 0 \end{pmatrix},$$

$$C_7 \equiv \begin{pmatrix} 1 & 0 & 0 \\ 0 & -1 & 0 \\ 0 & 0 & 0 \end{pmatrix},$$

$$C_8 \equiv \begin{pmatrix} 0 & 0 & 0 \\ 0 & 1 & 0 \\ 0 & 0 & -1 \end{pmatrix}.$$

6. Are they traceless?
7. Are they anti-Hermitian?
8. Pick A to be one of them.
9. Likewise, pick B to be one of them. (Later on, we'll extend this yet more.)
10. Span their commutator as

$$[A, B] = c_1C_1 + c_2C_2 + c_3C_3 + \cdots + c_8C_8,$$

for some real coefficients $c_1, c_2, c_3, \ldots, c_8$.
11. These coefficients are called structure constants.
12. They depend on A, B, and the original definition of our matrices $C_1, C_2, C_3, \ldots, C_8$.

16.7.4 *Strong Structure Constants*

1. So far, i and j indexed the quarks. Next, let $1 \leq i, j \leq 8$ index the matrices $C_1, C_2, C_3, \ldots, C_8$.
2. Let A be fixed and specific, as above: one of these eight matrices.
3. B, on the other hand, remains unspecified: either one of these eight matrices.
4. Define a new mapping: $[A, \cdot]$. What does it do? *Hint*: it maps B to $[A, B]$:

$$[A, \cdot] : B \to [A, B].$$

5. How does it look like? *Hint*: see the following.
6. Write it as an explicit 8×8 matrix, using $C_1, C_2, C_3, \ldots, C_8$ as a basis. *Hint*: to filter out its ith row, apply it to $B \equiv C_i$. This will filter out its $(i, \bar{j})$th element:

$$\begin{aligned} [A, \cdot]_{i,\bar{j}} &= [A, C_i]_j \\ &= c_j \\ &\equiv c_j\,(A, B \equiv C_i, C_1, C_2, C_3, \ldots, C_8) \end{aligned}$$

$(1 \leq i, j \leq 8)$. These are the strong structure constants. (This matrix works as in Section 16.5.1.)

7. Look at $-[A,\cdot]^*$. What is this physically? *Hint*: this is the antiperspective (Section 16.5.2).
8. Assume that gluon A attracts gluon B (strongly), by converting a red quark in A to a green quark in B, and vice versa (as in Figure 16.2 and Section 16.1.5). What is the probability for this? *Hint*: the probability for this is

$$|c_1|^2 + |c_2|^2$$

(assuming that both A and B are normalized to have sum-of-squares 1).
9. And what is the probability to do this by converting red to blue? *Hint*: $|c_3|^2 + |c_4|^2$.
10. And what is the probability to do this by converting green to blue? *Hint*: $|c_5|^2 + |c_6|^2$.

16.7.5 *Killing Form*

1. Look at a strong interaction: gluon A attracts gluon B.
2. For this, the ith gluon in B interchanges with the jth gluon in A (Figure 16.9).
3. What is the complex amplitude? *Hint*: the product:

$$[A,\cdot]_{i,\bar{j}}[B,\cdot]_{j,\bar{i}}.$$

4. But i and j are not yet specified. In fact, they could be just anything. Superpose over all possible i's and j's, and sum up:

$$\sum_{i,j=1}^{8}[A,\cdot]_{i,\bar{j}}[B,\cdot]_{j,\bar{i}} = \sum_{i=1}^{8}\left([A,\cdot]\ [B,\cdot]\right)_{i,\bar{i}} = \text{trace}\left([A,\cdot]\ [B,\cdot]\right).$$

5. What is this? *Hint*: this is a number, not a matrix.

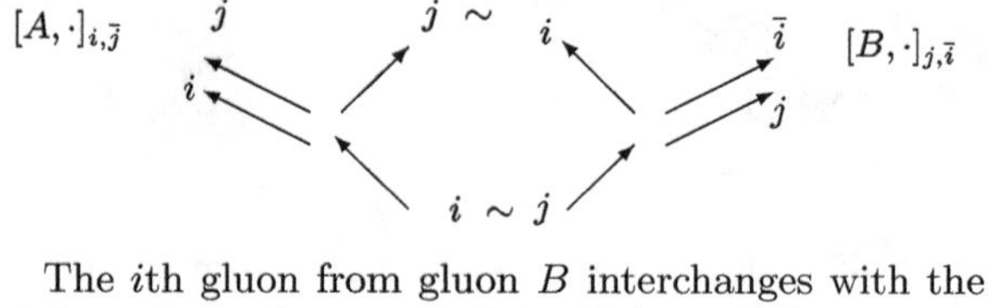

Figure 16.9. The ith gluon from gluon B interchanges with the jth gluon in gluon A. This is how gluon A attracts gluon B (strongly).

6. What is this physically? *Hint*: this is the complex amplitude that gluon A attracts gluon B (strongly) by exchanging an inner gluon.
7. Better yet, extend this yet more. For this, expand A and B linearly (in our basis):

$$A = \sum_{i=1}^{8} \alpha_i C_i, \quad \text{and} \quad B = \sum_{j=1}^{8} \beta_j C_j$$

(for some real coefficients α_i's and β_j's).
8. For these new A and B, what is the new complex amplitude? *Hint*: the new complex amplitude is now

$$\sum_{i,j=1}^{8} \alpha_i K_{i,j} \beta_j,$$

where K is the Killing form:

$$K_{i,j} = \text{trace}\left([C_i, \cdot] \ [C_j, \cdot]\right).$$

9. Use our basis for yet another important purpose: to write the commutator. *Hint*:

$$[A, B] = \sum_{i,j=1}^{8} \alpha_i \left[C_i, C_j\right] \beta_j.$$

10. What is this? *Hint*: this is a 3×3 matrix.
11. Why? *Hint*: this is just a linear combination of 3×3 matrices.

Chapter 17

Toward Quantum Field Theory

In quantum mechanics, we focus on one particle, or at most two entangled particles. In quantum field theory, on the other hand, we also ask: how was the particle born in the first place? And how could many more particles come and go? For this, we use operators: creation, annihilation, and counting. More precisely, these are density operators that can even be integrated. This gives us a complete framework to model how particles can even scatter from one another, and even spit a new particle, while preserving momentum and energy.

In this chapter, we go back to our active style: teaching new material through exercises. After reading each exercise, pause, and try to solve it on your own, before reading the hint (which is actually the solution). Enjoy!

17.1 Operators

17.1.1 *Position Operators*

1. Recall our original wave function: $u(x)$ (where $x \in \mathbb{R}$ is position, in 1-D). *Hint*: see Chapters 2 and 7.
2. Replace the complex number $u(x)$ by a new operator: $U^+(x)$.
3. What does U^+ do? *Hint*: it creates a new particle at x.
4. More precisely, U^+ is creation density (per unit length):

$$U^+(x) = \frac{d(\text{create a new particle at position} < x)}{dx}.$$

5. This is good for (1-D) integration:

$$\int_a^b U^+(x)dx = \text{create a new particle between } a \text{ and } b.$$

6. On the other hand, U^- drops an old particle from x.
7. More precisely, U^- is annihilation density (per unit length).
8. How to define it? *Hint*: for a particular x, this is the Hermitian adjoint:

$$U^- \equiv \left(U^+\right)^*.$$

9. For a particular x, look at their product: U^+U^-. What is this? *Hint*: this is the number operator that counts the particles at x.
10. More precisely, this is particle-number density (per unit length):

$$U^+U^-(x) = \frac{d(\text{number of particles at position} < x)}{dx}.$$

11. This is good for (1-D) integration:

$$\int_a^b U^+U^-(x)dx = \text{number of particles between } a \text{ and } b.$$

12. Look at $U^-(x)$ again. Write it as an $\mathbb{R} \times 1$ column vector:

$$\begin{pmatrix} \vdots \\ U^-(x) \\ \vdots \end{pmatrix}$$

(with x as a row index).
13. What are its components? *Hint*: each component is an operator (indexed by x): $U^-(x)$.
14. What is the norm of the entire vector? *Hint*:

$$\|U^-(x)\|^2 = \int_{-\infty}^{\infty} U^+U^-(x)dx = \text{total number of particles}.$$

17.1.2 Momentum Operators

1. Likewise, look at a new operator: $W^+(p)$ (or W^-, respectively). This operator creates (or drops) a particle of momentum p.
2. More precisely, this is creation (or annihilation) density (per unit momentum):

$$W^{+/-}(p) = \frac{d(\text{create/drop a particle of momentum} < p)}{dp}.$$

3. This is good for (1-D) integration:

$$\int_a^b W^{+/-}(p)dp = \text{create/drop a particle of } a < \text{momentum} < b.$$

4. For a particular p, how are W^+ and W^- related? *Hint*: Hermitian adjoint:

$$W^- = \left(W^+\right)^*.$$

5. Look at their product: $W^+W^-(p)$. What does it do? *Hint*: it counts the particles of momentum p.
6. More precisely, this is particle-number density (per unit momentum):

$$W^+W^-(p) = \frac{d(\text{number of particles of momentum} < p)}{dp}.$$

7. This is good for (1-D) integration:

$$\int_a^b W^+W^-(p)dp = \text{number of particles of } a < \text{momentum} < b.$$

8. For all p's together, W^- makes an $\mathbb{R} \times 1$ column vector:

$$\begin{pmatrix} \vdots \\ W^-(p) \\ \vdots \end{pmatrix}$$

(with p as a row index).
9. What are its components? *Hint*: each component is an operator (indexed by p): $W^-(p)$.

10. Likewise, interpret W^+ as an (infinite) row vector:

$$W^+ = \left(W^-\right)^* .$$

11. How are they related to $U^+(x)$ and $U^-(x)$, defined above? *Hint*: apply the Fourier transform to the entire column vector:

$$W^-(p) = F^*U^-(x),$$
$$W^+(p) = U^+(x)F.$$

12. Look at $W^-(p)$ again (as an infinite column vector). What is its norm? *Hint*:

$$\|W^-(p)\|^2 = \int_{-\infty}^{\infty} W^+W^-(p)dp = \text{total number of particles.}$$

13. Prove this in yet another way. *Hint*: the Fourier transform is unitary and preserves norm. Therefore,

$$\|U^-(x)\|^2 = \|W^-(p)\|^2 = \text{total number of particles.}$$

17.1.3 *Energy Operators*

1. Next, look at a fixed position: $x \in \mathbb{R}$. At this x, look at a new operator: $A^+(e)$ (or A^-, respectively). This operator creates (or drops) a particle of energy e.
2. More precisely, this is creation (or annihilation) density (per unit energy):

$$A^{+/-}(e) = \frac{d(\text{create/drop a particle of energy} < e)}{de}.$$

3. This is good for (1-D) integration:

$$\int_a^b A^{+/-}(e)de = \text{create/drop a particle of } a < \text{energy} < b.$$

4. For a particular e, what is their product? *Hint*: $A^+A^-(e)$ counts the particles of energy e.
5. More precisely, this is particle-number density (per unit energy):

$$A^+A^-(e) = \frac{d(\text{number of particles of energy} < e)}{de}.$$

6. This is good for (1-D) integration:

$$\int_a^b A^+A^-(e)de = \text{number of particles of } a < \text{energy} < b.$$

7. What are these particles? *Hint*: these could be electrons in the atom. This way, e is the energy level of the electron: $A^+(e)$ introduces an electron of energy e, $A^-(e)$ drops it, and $A^+A^-(e)$ counts it.
8. Still, this gives us little information about an individual electron, and its history. Where did it get its energy from? How did it get to this energy level?
9. How to improve on this model, and focus on just one electron? *Hint*: look at our electron as a complete physical system. Now, even in a general system, e can stand for all the energy, in the whole system. This way, each particle is not matter but energy: one quantum of energy. A^+ introduces one, A^- drops one, and A^+A^- counts one. Later on, we'll come back to this example.
10. Look at $A^-(e)$ again. Write it as an $\mathbb{R} \times 1$ column vector:

$$\begin{pmatrix} \vdots \\ A^-(e) \\ \vdots \end{pmatrix}$$

(with e as a row index).
11. What are its components? *Hint*: each component is an operator (indexed by e): $A^-(e)$.
12. What is the norm of the entire vector? *Hint*:

$$\|A^-(e)\|^2 = \int_{-\infty}^{\infty} A^+A^-(e)de = \text{total number of particles at } x.$$

17.1.4 *Fourier Transform and Radioactivity*

1. Apply the Fourier transform to the entire column vector. What do you get? *Hint*: the new column vector $(F^*A^-)(t)$ is annihilation density (per unit time).
2. What is this physically? *Hint*: absorption rate (into x).

3. Likewise, look at the (infinite) row vector A^+. Apply the Fourier transform to it (from the right). What do you get? *Hint*: the new row vector $(A^+F)(t)$ is creation density (per unit time).
4. What is this physically? *Hint*: emission rate (from x).
5. This is indeed radioactivity: particles emitted (or radiated) from x.
6. This is indeed uncertain (nondeterministic): you can never predict when the next particle will be emitted. All you can do is gather statistics and tell the half-life-time.

17.2 Photons and Energy Levels

17.2.1 *An Electron: A Complete System*

1. Let's look at the above example in some more detail.
2. For this, let the harmonic oscillator model just one electron in the atom. The electron is now the whole system: there are no more particles (of any matter).
3. Why is the harmonic oscillator suitable for this? *Hint*: unlike the massless photon, the electron has a positive mass: $m > 0$. Besides, let ω be its frequency in the atom.
4. In this system, what particle does A^+ create? *Hint*: a photon: one quantum of energy.
5. And what is e? *Hint*: see the following.
6. Should e be the energy of just one photon? *Hint*: this is not a very good idea because e could then take only two possible values: either $e = 0$, or $e = \omega\bar{h}$ (the energy in one photon). To count the photons, you'd then have to "integrate" vertically (and trivially): just count the photons in a tall tower above $e = \omega\bar{h}$.
7. Is there a better way? *Hint*: let e be the total energy in the whole system. Physically, this is much more meaningful and interesting: this is the energy level of the electron in the atom. If you like, this is the total number of photons "hidden" in the electron.
8. Why is this better? *Hint*: the photons are no longer concentrated vertically (in a tall tower above $e = \omega\bar{h}$), but distributed horizontally (and uniformly) over many (discrete) e-values: $e = \omega\bar{h}$, $2\omega\bar{h}$, $3\omega\bar{h}$, ... (one photon above each possible e-value).
9. How will A^+A^- count these photons? *Hint*: by integration over the entire e-axis (see the following).

10. Indeed, what is A^+A^- explicitly? *Hint*:

$$A^+A^- \equiv \frac{1}{\omega\bar{h}}.$$

11. Why must this be constant (independent of e)? *Hint*: the photons will be counted (or integrated, or summed up) horizontally: one by one, in a row (along the horizontal e-axis). In this summation, what is added is energy per energy (in the entire system), which must be constant.
12. In this model, what does integration mean physically? *Hint*: integration means gaining energy, to advance to a much higher energy level. From the ground state onward, each level adds one more photon, to get the electron to the next higher level.
13. For example, how many photons are needed to let the electron be in the nth energy level? *Hint*: n:

$$\begin{aligned}\int_{\omega\bar{h}/2}^{\omega\bar{h}(n+1/2)} A^+A^- de &= \int_{\omega\bar{h}/2}^{\omega\bar{h}(n+1/2)} \frac{1}{\omega\bar{h}} de \\ &= \frac{1}{\omega\bar{h}} \int_{\omega\bar{h}/2}^{\omega\bar{h}(n+1/2)} de = \frac{1}{\omega\bar{h}} n\omega\bar{h} = n.\end{aligned}$$

14. How does this integral produce a pure number (with no physical unit at all)? *Hint*: both de and $\omega\bar{h}$ share the same physical unit: energy.
15. For yet another example, to jump two levels ahead, how many photons must be absorbed? *Hint*: two:

$$\begin{aligned}\int_{\omega\bar{h}(n+1/2)}^{\omega\bar{h}(n+5/2)} A^+A^- de &= \int_{\omega\bar{h}(n+1/2)}^{\omega\bar{h}(n+5/2)} \frac{1}{\omega\bar{h}} de \\ &= \frac{1}{\omega\bar{h}} \int_{\omega\bar{h}(n+1/2)}^{\omega\bar{h}(n+5/2)} de = \frac{1}{\omega\bar{h}} 2\omega\bar{h} = 2.\end{aligned}$$

17.2.2 Absorbing a Photon

1. In this model, how does A^+ look like? *Hint*:

$$A^+ = \sqrt{\frac{m\omega}{2\bar{h}}} \left(X - \frac{i}{m\omega} P \right)$$

(Section 3.1.2).

2. What does it do physically? *Hint*: it produces a new photon. The electron absorbs it, gets stronger, and jumps ahead, to the next (higher) energy level.
3. Geometrically, how does A^+ look like? *Hint*: in the phase plane, it looks like an arrow (issuing from the origin, and pointing at $\bar{z}$: Section 3.4.2).
4. How does A^+ span $\mathbb{Z}$ (the additive group of integer numbers), to model anti-matter too? *Hint*: look at the ground state. Apply A^+ to it, time and again. This will produce $\bar{z}^n$. To model anti-matter too, look at inverse powers of z (Sections 3.4.2 and 4.3.3).
5. And what is A^-? *Hint*: this is the Hermitian adjoint of A^+. To obtain it, just replace i by $-i$ (since X and P are both Hermitian):

$$A^- = \left(A^+\right)^* = \sqrt{\frac{m\omega}{2\bar{h}}}\left(X + \frac{i}{m\omega}P\right).$$

6. Why is this the Hermitian adjoint? *Hint*: X and P are both Hermitian.
7. What does A^- do physically? *Hint*: the electron spits a photon and falls to the lower energy level.

17.3 Beam of Light

17.3.1 *Beam of Light*

1. So far, we used the harmonic oscillator to model an electron in the atom. The electron was the whole system. The photons helped it "climb" to higher and higher energy levels.
2. Next, look at a new system: a laser beam, containing n (identical) photons (where n is nondeterministic, and ω is the color).
3. Can this be modeled by the harmonic oscillator? *Hint*: no! The photon is massless: $m = 0$.
4. In a photon, how is the frequency ω related to the wave number k? *Hint*: as in a wave:

$$\omega = ck$$

(where c is the speed of light).
5. Is this consistent in terms of physical unit? *Hint*: both sides have the same physical unit: 1/time (since k has unit 1/length).

6. Multiply both sides by $\bar{h}$. What do you get? *Hint*:

$$\omega\bar{h} = ck\bar{h} = cp$$

(Sections 2.2.1 and 2.5.5).
7. What does this mean physically? *Hint*: in our laser beam, this is how the energy of each photon is related to its momentum.
8. Multiply both sides by n (the total number of photons). What do you get? *Hint*:

$$H = n\omega\bar{h} = ncp = cnp = cP.$$

9. What does this mean physically? *Hint*: the Hamiltonian (that measures nondeterministic energy) depends on the momentum operator (that measures nondeterministic momentum). They actually make the same random variable, with the same uncertainty in it.
10. Where did this uncertainty come from? *Hint*: recall that n is nondeterministic.
11. Look at the above equation again. What does it tell us algebraically? *Hint*: in our laser beam, P commutes with H.
12. What does this tell us physically? *Hint*: not only energy but also momentum is conserved.
13. Is this different from the harmonic oscillator? *Hint*: in the harmonic oscillator, energy depends on position too, which doesn't commute with momentum. For this reason, H and P don't commute. Therefore, only energy is conserved but momentum is not.

17.4 On Scattering

17.4.1 *On Scattering*

1. So far, U^- was a column vector, and U^+ a row vector.
2. Here, on the other hand, make a little change, and look at things the other way around: interpret U^- as a row vector, and U^+ as a column vector.
3. This will help design a new matrix, whose elements are not numbers but operators (say, particle-count operators).

4. For example, Dirac's delta function will be interpreted here with operators on its main diagonal (see the following).
5. For yet another example, look at the new density matrix (for position alone). Its elements are now operators in their own right:
$$U^+(x)U^-(\tilde{x})$$
(where x is the row index, and $\tilde{x}$ the column index).
6. In particular, look at a special case, in which this is diagonal:
$$U^+(x)U^-(\tilde{x}) = \delta(x - \tilde{x})$$
(a diagonal matrix, with the same operator U^+U^- along its main diagonal).
7. Write this as an explicit diagonal matrix. *Hint*:
$$\operatorname{diag}\left(U^+U^-(x)\right)$$
(where x is the diagonal index, and $\tilde{x}$ is gone).
8. What is this physically? *Hint*: this models scattering. A static target is placed at some (fixed) position x. (We don't care where x is.) A particle hits it, gets absorbed, and reemitted back again (instantaneously).
9. Is this physical? *Hint*: not quite. This is a bit too ideal. In real physics, the emission occurs a little later. Still, let's disregard this delay.
10. Is it reemitted with the same momentum? *Hint*: see the following.
11. Write yet another density matrix (for momentum, not position). *Hint*: apply the Fourier transform (from both sides). Don't forget our little change: W^+ is now a column vector, and W^- a row vector:
$$\begin{aligned} W^+(p)W^-(\tilde{p}) &= F^t U^+(x)U^-(\tilde{x})\left(F^*\right)^t \\ &= F^t\delta(x - \tilde{x})\left(F^*\right)^t = \delta(p - \tilde{p}) \end{aligned}$$
(a new diagonal matrix, with the operator W^+W^- along its main diagonal).
12. Write this as an explicit diagonal matrix. *Hint*:
$$\operatorname{diag}\left(W^+W^-(p)\right)$$
(where p is the new diagonal index and $\tilde{p}$ is gone).

13. What does this mean physically? *Hint*: conservation of momentum. The particle gets reemitted with the same momentum:

$$p = \tilde{p}.$$

14. Is this physical? *Hint*: not quite. In real physics, the target also moves a little, and takes a little bit of incoming momentum, and doesn't give it back at all. Still, let's disregard this (tiny) loss.
15. This is single emission. (Just one particle is emitted.) How about double emission (two particles are emitted from the same point x)? *Hint*: see the following.

17.4.2 Double Emission

1. For this, we'll need to replace the operator W^+ by its square: $(W^+)^2$. How? *Hint*: see the following.
2. In the above, replace $-$ by $+$, to obtain a new skew-diagonal matrix [with $(W^+)^2$ along its skew-diagonal]. *Hint*:

$$W^+(p)W^+(\tilde{p}) = F^tU^+(x)U^+(\tilde{x})F = F^t\delta(x - \tilde{x})F = \delta(p + \tilde{p})$$

(since both emissions come from the same point: $x = \tilde{x}$).
3. In the latter matrix, what do you have along the skew-diagonal? *Hint*: the operator

$$W^+(p)W^+(\tilde{p}), \quad \text{if } p = -\tilde{p}.$$

4. And what do you have elsewhere? *Hint*: the zero operator (if $p \neq -\tilde{p}$).
5. What is this physically? *Hint*: double emission (from the same position), with no absorption at all.
6. To model emission only, interpret the latter matrix as a column vector. *Hint*: see the following.
7. To do this, rename

$$a \equiv p, \quad \text{and} \quad b \equiv \tilde{p}.$$

8. Next, introduce a new row index:

$$p \equiv a + b$$

(Figure 17.1).

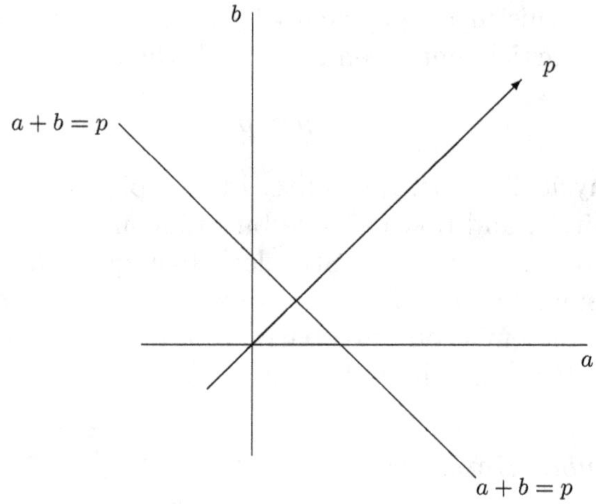

Figure 17.1. Double emission: emitting two particles (of momenta a and b). Look at those pairs (a, b) that sum to $a + b = p$ (a fixed outgoing momentum). All such pairs are as likely, and should be superposed (integrated with respect to db). This will form a new emission operator (the pth component in our new column vector).

9. Does this agree with what we did in the beginning? *Hint*: here, there is no absorption (no incoming momentum at all):

$$\tilde{p} = 0.$$

To agree with what we did in the beginning, we must have

$$\delta(a + b) = \delta(p - \tilde{p}) = \delta(p - 0) = \delta(p),$$

or indeed

$$p = a + b.$$

10. As before, this delta function returns an operator, not a number.
11. Sum the skew-diagonal. What do you get? *Hint*: a superposition over all those pairs (a, b) that sum to zero:

$$a + b = 0.$$

After all, all these pairs are as likely.
12. For now, this delta function is centered at 0 (no incoming momentum at all). Later on, we'll shift it, and center it at $\tilde{p}$ back again, to welcome any (nonzero) momentum as well. In this case,

how will the skew-diagonal sum? *Hint*: it will be a superposition over all those pairs (a, b) that sum to $\tilde{p}$:

$$a + b = \tilde{p}.$$

13. This will sum operators, not numbers. What operator exactly? *Hint*: double-emission density, producing total momentum $\tilde{p}$ (see the following).
14. To see this better, write the outcome as a new column vector (with p as a new row index). What is its pth component? *Hint*: double-emission density, producing total momentum p (see the following).
15. To have this, superpose over all those pairs (a, b) that sum to p. (After all, all these pairs are as likely.) Design a new column vector:

$$\begin{pmatrix} \vdots \\ \int_{-\infty}^{\infty} W^+(p-b)W^+(b)db \\ \vdots \end{pmatrix}$$

(where p is the row index).
16. This integrates on db. Why? *Hint*: $W^+(b)$ is emission density at b (per unit momentum), which must be multiplied by the momentum element: db.
17. But something is still missing: absorption. To model this too, multiply by $W^-(\tilde{p})$ from the right (to welcome momentum $\tilde{p}$ first). As we saw in the beginning, this will produce a diagonal density matrix (for momentum alone):

$$\int_{-\infty}^{\infty} W^+(p-b)W^+(b)W^-(\tilde{p})db = \delta(p-\tilde{p})$$

(where p is the row index and $\tilde{p}$ is the column index).
18. Write this more explicitly: a diagonal matrix, with the same operator along its main diagonal:

$$\text{diag}\left(\int_{-\infty}^{\infty} W^+(p-b)W^+(b)W^-(p)db\right)$$

(where p is the diagonal index, and $\tilde{p}$ is gone).

19. What is this physically? *Hint*: conservation of momentum.
20. Is this physical? *Hint*: not quite. This is a bit too ideal. In real physics, the target moves a little, and takes a little bit of incoming momentum, never giving it back any more. This way, the emitted particles lose a little bit of momentum for good.
21. Look at things from the perspective of the target itself. What does it see? It sees an incoming particle (with a particular momentum), and a pair of new outgoing particles (with a little less total momentum). How come? What about conservation of momentum? *Hint*: the target on its own is an open system that welcomes force, energy, and momentum from the outside. Likewise, the particle on its own is an open system that leaks force, energy, and momentum to the outside. Only together are they a closed (isolated) system, with conservation laws.
22. In the above, replace x by t (time) and p by e (energy). Obtain a similar conservation law for energy as well.

Part VII: Algorithms in C++

What is a quantum computer? This is a k-dimensional hypercube, with 2^k binary nodes. On it, the joint (discrete) wave function is actually a grid function: the complex amplitude (of each and every node). Thanks to entanglement, this is a genuine (nonseparable) grid function, with $2^k - 1$ degrees of freedom: 2^k nodes, minus one (for normalization). This is the great power of quantum computing: exponential in k.

Better yet, model it as a tensor, with k indices. Each index is binary: either 0 or 1. This makes a qubit. Together, our k qubits make our quantum computer: 2^k different configurations. On each, the tensor is defined in a different (and independent) way: the complex amplitude. This is entanglement again: nonseparability: the tensor can never be decomposed as a product of smaller tensors (Section 8.3.2). This is why it has $2^k - 1$ degrees of freedom, no less.

Why is the tensor better? Well, thanks to it, our quantum computer gets more general: not only binary but also N-ary. In fact, $N = 2$ models spin, as usual. (With $k = 2$, for instance, our tensor is just a 2×2 matrix, as in a boson.) But this is not the only way: in theory, $N = 3$ could model a gluon (Section 16.7.3), and so on.

This is still quite theoretical. After all, nobody built a quantum computer yet. Still, there is a hope. Technology progresses quite fast: you may wake up one morning, and find a new quantum computer on your desk! Better get ready: develop quantum algorithms, and implement them on your good old digital computer for now. How? Best do this in an object-oriented programming language like C++. This way, you can program a new class: a dynamic tree, with as many levels as you like.

In Chapter 6, we already saw a fundamental structure: the binary tree, useful in recursive algorithms. In a binary tree, each node has two branches: either leftward or rightward. This is indeed the most elementary ingredient in information theory: a bit, which can take two possible values: either 0 or 1. This is indeed the basis for the digital computer. In a quantum computer, on the other hand, the (little) bit grows and develops into a new random variable: a qubit, which could be either 0 (at some probability) or 1 (at some other probability: Chapters 13–15).

The binary tree has a more general version: an N-ary tree. In it, each node has not only two but N branches. (In relativity, for example, we often use $N = 4$, to model a four-dimensional differentiable manifold: spacetime.) Thanks to this, we can even implement dynamic tensors (with as many indices as you like) and their arithmetic operations.

How to do this in practice? Here, we offer a C++ code, and explain it in detail. It is based on the framework in Ref. [24]. Still, it stands on its own legs. This is indeed the power of an object-oriented language: it lets you design a new class, based on older classes, as in math.

Finally, we also show how to use C++ in a practical problem from computational electromagnetics. On a mesh of finite elements, linearize the Maxwell system, and assemble the stiffness-mass matrix. To understand the C++ code in detail, better look at Ref. [24] first. Or, if you like, read the present code on its own. After all, thanks to the detailed explanation, it stands on its own legs. In fact, the idea is quite simple: scan the finite elements, one by one, in a row, and assemble the stiffness and mass matrices on each. The stiffness matrix discretizes the derivatives: the curl of the electric field E (the unknown field in the system). The mass matrix, on the other hand, is a bit trickier: it also contains a nonlinear term, which must be linearized time and again (at each and every Newton iteration, at the up-to-date E). Once the linear system is solved numerically (at the present iteration), the new E can help relinearize and get ready for the following iteration.

Chapter 18

Dynamic Tensors in C++

Thanks to information theory, we already have a useful mathematical structure: the binary tree. In it, each node has two branches: leftward or rightward. At its bottom, however, the binary tree has a different kind of nodes: leaves, with no branches at all.

The bottom level could be mirrored by a new geometrical structure: a hypercube. Each leaf in the original tree is mirrored by a corner in the hypercube. We already used this to design our quantum computer.

How to mirror? Use mathematical induction! After all, as you jump from level to level, the number of leaves doubles. Likewise, as the hypercube grows by one dimension, the number of corners doubles. Likewise, as your tensor grows by one index, the number of entries doubles. What is so good about the tensor? Well, it is not only geometrical but also algebraic: ready for all sorts of arithmetic operations.

This is true for not only a binary but also an N-ary tree. In it, each node has not only two but N branches. This makes a bigger tensor: in each dimension, it takes now more possible numbers: not only 0 or 1 but also 2 or 3 or $4, \ldots,$ or even $N - 1$. In relativity, for example, we often use $N = 4$, to let an index be either 0 (time) or 1 or 2 or 3 (space). This helps span spacetime, and design its metric.

In this chapter, we use an N-ary tree to implement our tensor, with all sorts of arithmetic operations. Thanks to this, our tensor gets highly dynamic: as big as you like, with as many indices as you like.

How to do this on the computer? Best use an object-oriented language like C++. This way, the code follows the math: addition, multiplication, and even contraction. In your N-ary tree, use as many levels as you like. This way, your tensor will have as many indices as you like.

18.1 Dynamic N-ary Tree

18.1.1 *Dynamic Tensor*

To get started, look at a small tensor, with just two indices: m and n. Actually, this is a matrix. In a boson (where $N = 2$), this is a 2×2 matrix. In a gluon (where $N = 3$), this is a 3×3 matrix. In relativity (where $N = 4$), this is a 4×4 matrix (indexed by $m, n = 0, 1, 2, 3$). Later on, we'll extend this to a bigger tensor: a $4 \times 4 \times 4$ tensor, indexed by $\mu, \nu, \alpha = 0, 1, 2, 3$. Higher yet, we'll even use a bigger tensor (of 4^4 entries), indexed by $\rho, \mu, \sigma, \nu = 0, 1, 2, 3$, and so on.

How to implement all these in one go? Best use dynamic tensors, with as many indices as you like. This way, you can write your code once and for all.

18.1.2 *Full N-ary Tree*

For this purpose, we need a full tree: each level has N subtrees, where N is still unspecified. (Our default will be $N = 4$.) How many levels are there? As many as the number of indices in the tensor. This way, each leaf will store one entry. In Figure 18.1, for instance, we use two levels to store a matrix (of 16 entries).

What is the individual entry? In C++, it could be just anything. For this, we denote it by T: an implicit type (to be specified later as a complex number, or even a polynomial). We only need T to have arithmetic operations like addition and multiplication.

18.1.3 *The Tensor Class*

First, we define the tensor class. To store data, it will have two fields. The first field is called "entry". (These standard quotation marks tell you that this word will actually be used in the code.) In our tree, at the bottom level, in each leaf, "entry" will store an entry. At higher

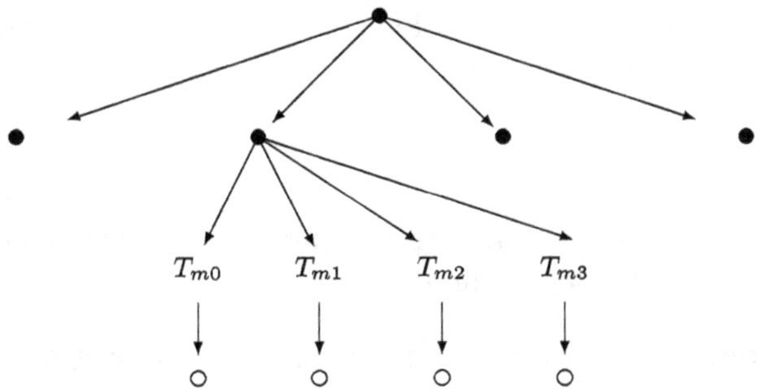

Figure 18.1. A 4×4 tensor, implemented as a two-level tree. From each node, four branches issue, pointing downward (toward the next lower level). At the bottom, the leaves have no branches but only entries: T_{mn} $(0 \leq m, n \leq 3)$.

levels, on the other hand, the nodes will store no entry at all: their "entry" field will remain empty.

The second field, "next", will point to N subtensors (to be allocated memory later on):

```
template<class T, int N> class tensor{
  protected:
    T entry;
    tensor* next[N];
  public:
    tensor(const T&s=0, tensor* p=0):entry(s){
      for(int i=0; i<N; i++)
        next[i] = p;
    } //  default constructor
```

18.2 Constructors and Destructor

18.2.1 *Default Constructor*

The above constructor takes two arguments: "s" (to serve as the entry) and "p" (to point to a subtensor). If the constructor is called with no arguments at all, then it would use the default values: zero entry, and no subtensors at all. This would make a very small "tree": a dangling leaf, ready to be placed in a bigger tree later on.

If, on the other hand, the constructor is called with some concrete arguments "s" and "p", then all subtensors would coincide and duplicate. This would make little sense and should be avoided. Let's move on to a better constructor.

18.2.2 *Copy Constructor*

The copy constructor, on the other hand, takes another kind of argument: an existing tensor, called "t":

```
tensor(const tensor&t):entry(t.entry){
  for(int i=0; i<N; i++)
    if(t.next[i])
```

From "t", copy each subtensor (recursively):

```
        next[i] = new tensor(*t.next[i]);
      else
        next[i] = 0;
} //  copy constructor
```

18.2.3 *Destructor*

How to remove an old tensor? For this, how to remove its subtrees and release their memory? Recursively (by address):

```
~tensor(){
  for(int i=0; i<N; i++){
    delete next[i];
    next[i] = 0;
  }
} //  destructor
```

18.3 Member Functions

18.3.1 *Number of Levels*

How many levels are there? Or, in our tensor, how many indices are there? Use recursion again:

```
int indices() const{
  return next[0] ? next[0]->indices() + 1 : 0;
} //  total number of indices
```

18.3.2 *Read an Entry*

In our tree, only the leaves store genuine entries. How to read them? Like this:

```
const T& operator()() const{
  return entry;;
} //  read an entry
```

Thanks to the word "const", this function can never be used to change an entry.

18.3.3 *Read a Subtensor*

How to read a subtensor? Either by

```
const tensor* readNext(int i) const{
  return next[i];
} //  read ith subtensor
```

(read only) or by

```
tensor* Next(int i){
  return next[i];
} //  read/write ith subtensor
```

(read/write). Next, declare some more member functions:

```
const tensor& operator=(const tensor&);

const tensor& operator+=(const tensor&);

const tensor& operator*=(const T&);

const tensor operator*(const T&) const;

const tensor& operator*=(const tensor&);

void replace(int, int);

void contractFirst(int);
```

```
    void contract(int, int);
};
```

18.4 Member Operators

18.4.1 *Assignment Operator*

How to define them? Here is the assignment operator:

```
template<class T, int N>
const tensor<T,N>&
tensor<T,N>::operator=(const tensor<T,N>&t){
  if(this != &t){
    entry = t.entry;
    for(int i=0; i<N; i++)
      if(next[i])
        if(t.next[i])
```

If there are old subtensors, then update them (recursively):

```
          *next[i] = *t.next[i];
        else{
          delete next[i];
          next[i] = 0;
        }
      else
        if(t.next[i])
```

Otherwise, copy them:

```
          next[i] = new tensor(*t.next[i]);
  }
  return *this;
} //  assignment operator
```

18.4.2 *Addition*

How to add tensors? Recursively again:

```
template<class T, int N>
const tensor<T,N>&
tensor<T,N>::operator+=(const tensor<T,N>&t){
```

Here, the argument "t" is added to the current tensor. For this, they must have the same number of indices. (Otherwise, issue an error message, which is left as an exercise.) If there are no indices at all (no subtrees), then add the dangling leaf:

```
    if(!next[0])
      entry += t.entry;
```

Otherwise, add the subtensors (recursively), one by one:

```
    else
      for(int i=0; i<N; i++)
        *next[i] += *t.next[i];
    return *this;
  } //  add a tensor
```

18.5 Multiplication

18.5.1 *Tensor Times Scalar*

How to multiply by a scalar "s"? Recursively again:

```
  template<class T, int N>
  const tensor<T,N>&
  tensor<T,N>::operator*=(const T&s){
    if(!next[0])
      entry *= s;
```

Indeed, multiply the subtensors recursively, one by one:

```
    else
      for(int i=0; i<N; i++)
        *next[i] *= s;
    return *this;
  } //  multiply by a scalar
```

Thanks to this, you can now write "Q *= q" to multiply the tensor "Q" by the scalar "q", and store the result back in "Q". Still, how to write "Q * q" (without changing "Q" at all)? For this, we must have yet another operator:

```
  template<class T, int N>
  const tensor<T,N>
```

```
tensor<T,N>::operator*(const T&s) const{
  return tensor(*this) *= s;
} //  tensor times scalar
```

18.5.2 *Tensor Times Tensor*

Thanks to this, we can now implement tensor-times-tensor as well:

```
template<class T, int N>
const tensor<T,N>&
tensor<T,N>::operator*=(const tensor<T,N>&t){
```

If the current tensor is just a dangling leaf (no subtrees at all), then use it to multiply "t":

```
    if(!next[0])
      *this = t * entry;
```

Otherwise, scan the current subtensors, one by one:

```
    else
      for(int i=0; i<N; i++)
```

and multiply each by "t" (recursively):

```
        *next[i] *= t;
    return *this;
} //  multiply by another tensor
```

This way, in the innermost recursive call, all leaves will be multiplied by "t", as required. The result will be the tensor product: a new (big) tensor, containing all possible products of an entry from the original tensor times an entry from "t".

18.6 Contraction

18.6.1 *Replace a Tree by a Subtree*

Look at the Riemann tensor $R^{\rho}_{\mu\sigma\nu}$. It has four indices: ρ, μ, σ, ν. Let's contract ρ with σ, to yield the Ricci tensor: $R_{\mu\nu}$. For this purpose, we don't need all 16 instances $0 \leq \rho, \sigma \leq 3$, but only the four diagonal

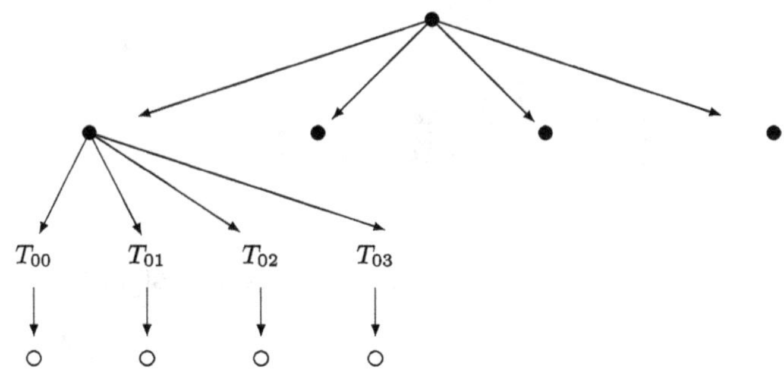

Figure 18.2. In the "replace" function, the original 4×4 tensor $T_{\mu\nu}$ (which contains 16 entries) is replaced by its leftmost subtree (which contains four entries only).

ones: $\sigma = \rho = 0, 1, 2, 3$. In the contraction, these will be summed up (as in a trace). For this purpose, σ is no longer independent but depends on ρ, and must be the same. In fact, σ can drop out, leaving only those σ's that are the same as ρ.

In the Riemann tensor, σ is the third index. This makes our task a bit difficult. Let's start from an easier task.

Consider a 4×4 tensor $T_{\mu\nu}$. Suppose that we need only four entries in it: say, $T_{0\nu}$ $(0 \le \nu \le 3)$. This is the leftmost subtree in Figure 18.2.

So, we actually don't need μ any more. We can safely drop it, and obtain a new smaller tensor, with one index only. To do this, we need to replace the original tree by its own (leftmost) subtree:

```
template<class T, int N>
void
tensor<T,N>::replace(int pick, int index){
```

Here, there are two arguments: "index" tells us which index is dropped. (In the above example, μ is the first index in T, so "index"= 1.) Moreover, the argument "pick" tells us which son to pick. In the above example, we only want to pick those entries with $\mu = 0$, so "pick"= 0 (the firstborn son).

So, in our example, we only need to keep the leftmost subtree. (Firstborn son replaces father.) The other three, on the other hand, may safely drop:

```
  if(index==1){
    entry = next[pick]->entry;
```

```
      for(int i=0; i<N; i++)
        if(i != pick){
          delete next[i];
          next[i] = next[pick]->next[i];
        }
```

By now, four items were copied from the leftmost subtree into the current tree:

- Its head (the individual son replaced his own individual father) and
- three subtrees: three grandsons replaced their own uncles (respectively). This was done by address, to avoid copy big T objects.

There is one more subtree to copy (the fourth grandson should replace his own father):

```
      next[pick]->entry.~T();
      next[pick] = next[pick]->next[pick];
    }
```

But what about our original task? In the Riemann tensor, σ is the third index, so the present function should actually be called with "index"$= 3$, not 1. For this, we need recursion:

```
    else
     for(int i=0; i<N; i++)
       next[i]->replace(pick, index-1);
  } //  replace the current tensor by the
          "pick"th subtensor
```

18.6.2 *Contract with the First Index*

This will help complete our original task: in the Riemann tensor $R^{\rho}_{\mu\sigma\nu}$, contract the first index ρ with the third index σ:

```
  template<class T, int N>
  void
  tensor<T,N>::contractFirst(int index){
```

Indeed, this function will contract with the first index: ρ. To contract ρ with σ, call it with "index" $= 3$. This way, σ will drop and will take the dependent value $\sigma = \rho =$"i":

```
    for(int i=0; i<N; i++)
      next[i]->replace(i, index-1);
```

To contract with ρ, just sum the subtensors and store the sum in the leftmost subtree:

```
    for(int i=1; i<N; i++)
      *next[0] += *next[i];
```

Finally, put this in the head:

```
    replace(0, 1);
  } //  contract with the first index
```

18.6.3 *Contract with Any Index*

Our original task is now complete: contract with the first index: ρ. Still, what about a more difficult task: say, contract the second and fourth indices with each other? For this purpose, call the next function (with "index1"$= 2$ and "index2"$= 4$):

```
  template<class T, int N>
  void
  tensor<T,N>::contract(int index1, int index2){
    if(index1 == 1)
```

If "index1"$= 1$, then this should contract with the first index (we already know how):

```
      contractFirst(index2);
    else
```

If, on the other hand, "index1"> 1, then use recursion:

```
      for(int i=0; i<N; i++)
        next[i]->contract(index1-1, index2-1);
  } //  contract
```

18.6.4 *Print the Tensor*

Finally, define an ordinary (nonmember) function to print the entries on the screen:

```
template<class T, int N>
void print(const tensor<T,N>& t){
```

To increase efficiency, the function takes the tensor "t" not by name but by a (constant) reference. If "t" is just a leaf, then print it:

```
    if(!t.readNext(0))
      print(t());
```

Otherwise, use recursion:

```
    else
      for(int i=0; i<N; i++)
        print(*t.readNext(i));
} //  print the leaves
```

Chapter 19

Maxwell's Equations: Linearize and Assemble on Finite Elements

In this chapter, we implement an important algorithm in computational physics: how to solve a nonlinear system of PDEs, more specifically, how to linearize and assemble the Maxwell equations on (linear) finite elements. The logic is quite simple: scan the finite elements, one by one. On each, assemble the stiffness and mass matrices (using a nested loop). The stiffness matrix discretizes the derivatives: the curl of the (unknown) electric field E. The mass matrix, on the other hand, is a bit trickier: in each Newton iteration (that converges to E), it must be recalculated and relinearized around the up-to-date E. Once the linear system is solved numerically, E gets updated, ready for the following iteration.

The code is based on Chapter 29 in Ref. [24]. Still, it improves on it in two aspects:

- It discretizes the curl directly. This is ready to extend to dielectric materials as well.
- The PDEs (including the nonlinear term) are defined (and linearized) in 3-D. This is ready to extend to other systems as well.

19.1 Maxwell's Equations

19.1.1 The Maxwell Equations

How do the Maxwell equations look like? Originally, they integrate in a 3-D domain. The left-hand side integrates in the entire domain, whereas the right-hand side integrates on some boundary (say, on the left edge of the domain):

$$\iiint \left((\nabla \times E)^t (\nabla \times \tilde{E}) + K E^t \tilde{E} + K_1 \|E\|^2 E^t \tilde{E} \right) dxdydz$$
$$= \sqrt{K} \iint (\mathbf{n} \times B)^t \, \tilde{E} dxdy,$$

where K and K_1 are given (negative) parameters, E is the (unknown) electric field, $\tilde{E}$ is a test vector function, B is the (unknown) magnetic field, and $\mathbf{n}$ is the outer normal vector (perpendicular to the boundary).

19.1.2 The Finite-Element Mesh

In our domain, embed a mesh of tetrahedra. This way, E will be approximated as a linear combination of (linear) nodal basis functions ψ_J (where J indexes the mesh nodes), with unknown (complex) coefficients c_J:

$$E \doteq \begin{pmatrix} \sum_J c_J^{(1)} \psi_J \\ \sum_J c_J^{(2)} \psi_J \\ \sum_J c_J^{(3)} \psi_J \end{pmatrix}.$$

The test function $\tilde{E}$, on the other hand, could be just any nodal basis function:

$$\tilde{E} = \begin{pmatrix} \psi_I \\ 0 \\ 0 \end{pmatrix} \quad \text{or} \quad \begin{pmatrix} 0 \\ \psi_I \\ 0 \end{pmatrix} \quad \text{or} \quad \begin{pmatrix} 0 \\ 0 \\ \psi_I \end{pmatrix}$$

(where I indexes the mesh nodes). Now, on our mesh, how to linearize and assemble?

19.1.3 *The Stiffness and Mass Matrices*

On our mesh, we'll have a discrete system. Its coefficient matrix will be the sum of two matrices:

- The stiffness matrix will discretize the curl: $\nabla \times E$ (where E is the unknown electric field).
- The mass matrix, on the other hand, will contain two terms:
  - the linear term (with the coefficient K, stored in the external parameter "HELM") and
  - the nonlinear term (with the coefficient K_1, stored in "HELM-Nonlin").

In our numerical examples, we often pick $K = K_1 = -20$.

19.2 How to Linearize?

19.2.1 *Linearization*

Let's start from the latter term: the nonlinear term. "Thanks" to it, our discrete system will get nonlinear:

$$A(\tilde{x}) = \tilde{f}.$$

Here, $A()$ is a nonlinear mapping that maps $\tilde{x}$ (the unknown numerical approximation to E) to $\tilde{f}$ (the given discrete right-hand side). This will yield the new (linearized) system

$$Ae = f.$$

How? Suppose that we already have a good approximation

$$x \doteq \tilde{x}.$$

Still, how to improve on x yet more? For this, we need to approximate its error:

$$e \equiv e(x) \equiv x - \tilde{x}.$$

For this, look at its residual:

$$f \equiv f(x) \equiv A(x) - \tilde{f}.$$

Moreover, let A be the Jacobian at x:

$$A \equiv A'(x).$$

We are now ready to solve for e (approximately):

$$Ae = f.$$

In this (linear) system, the coefficient matrix A will be designed in our constructor: a new C++ function, which will take three arguments:

- the mesh "m",
- the vector "x": a grid function (defined on the mesh nodes), ready to store x (the current approximation to $\tilde{x}$), and
- the vector "f", ready to store its residual:

$$f \equiv f(x) \equiv A(x) - \tilde{f}.$$

(These standard quotation marks tell us that m, x, and f belong in our C++ code.)

19.2.2 *Newton's Iteration*

In each Newton iteration, improve on x yet more. For this, substitute

$$x \leftarrow x - e.$$

How to get ready for the next iteration as well? At this new x, recalculate

$$A \leftarrow A'(x), \quad \text{and} \quad f \leftarrow f(x).$$

(This will be done in the same constructor as well.) This gets you ready for the next iteration as well. Better yet, in A, no need to recalculate the stiffness matrix, which is linear, and independent of x. Leave it as is, and only update the nonlinear term: relinearize it at the new x.

19.3 The Implementation

19.3.1 *The Coefficient Matrix and Its Elements*

In our new constructor, A will be the current matrix (that is currently being constructed). In each Newton iteration, x will get improved.

Thanks to this new x, the constructor can be called again, to recalculate the nonlinear term in f, and relinearize it in A. This way, both A and f get ready for the next iteration as well.

In A, the elements are of type T, which could be just anything. Later on, T will get specified as "matrix2": a little 2×2 matrix, ready to store the (complex) nonlinear term, linearized in real arithmetic: its real and imaginary parts are both differentiable and can get linearized separately.

19.3.2 *The Initial Guess*

Newton's iteration will converge to the numerical solution: h$\tilde{x}$. How to start? Pick an initial guess "x", with its initial residual:

$$\text{"f"} = A(\text{"x"}) - \tilde{f}.$$

For example, "x" could be picked as the zero vector. Often, $A(\mathbf{0}) = \mathbf{0}$, so "f" is initially $-\tilde{f}$.

19.3.3 *Vectors and Their Components*

In Newton's iteration, we'll improve on "x":

$$x \leftarrow x - e,$$

until it approximates $\tilde{x}$ well enough. Here, both "x" and "f" are grid functions, defined on the mesh nodes. On each node, "x" actually contains three scalars, to form the electric field E at this point. Each scalar is actually a complex number, stored as a pair: the real part, followed by the imaginary part. Later on, this will help differentiate in real arithmetic.

19.3.4 *The Mesh Nodes*

We use linear finite elements: each tetrahedron contains four degrees of freedom (one at each corner). In the mesh "m", the total number of nodes will be stored in the parameter "nodes" (to be specified later). Since the electric field contains three spatial components, A has a 3×3 block form. Each block is of order 2 times "nodes" (to store the real and imaginary parts).

19.4 The Constructor of the Matrix

19.4.1 *The Constructor*

To produce the matrix A, the constructor takes three arguments: the mesh "m" and the vectors "x" and "f":

```
template<class T>
sparseMatrix<T>::sparseMatrix(mesh<tetrahedron>&m,
    const runVector<point>&x,
    runVector<point>&f){
  int nodes = m.indexing();
```

The function "indexing" indexes the mesh nodes and returns their total number [24]. The boundary conditions will be of mixed-Neumann type. Where to impose mixed conditions? Where the vector "xBoundary" is nonzero. (In our example, this will be at the left edge of the domain.)

```
  runVector<int> xBoundary(nodes,0);
```

19.4.2 *Initialization to Zero*

In the beginning, the matrix contains no rows at all:

```
  this->item = new row<T>*[this->number = 3 * nodes];
  for(int i=0; i<this->number; i++)
    this->item[i] = 0;
```

Assume that we already updated "x" (in some Newton iteration). How to get ready for the next iteration as well? For this, we must relinearize: recalculate A at the up-to-date "x". Besides, we also need to recalculate the residual:

$$\text{“f”} = A(\text{“x”}) - \tilde{f}.$$

For this purpose, initialize "f" as the zero (complex) vector:

```
  f = point(0.,0.);
```

19.5 The Unit Tetrahedron

19.5.1 *Nodal Functions*

In the unit tetrahedron T (Figure 19.1), define four nodal functions: p_0, p_1, p_2, and p_3. Each is a linear polynomial (of three variables): 1 at one corner and 0 at the other three. To design them, we need a few polynomials of one variable (named with a low-case letter) or two variables (with a capital letter):

```
polynomial<double> zero(1,0.);
polynomial<polynomial<double> > Zero(1,zero);
polynomial<polynomial<polynomial<double> > >
    ZZero(1,Zero);
polynomial<double> one(1,1.);
polynomial<polynomial<double> > One(1,one);
polynomial<double> minus1(1,-1.);
polynomial<polynomial<double> > Minus1(1,minus1);
polynomial<double> oneMinusx(1.,-1.);
polynomial<polynomial<double> >
    oneMinusxMinusy(oneMinusx,minus1);
polynomial<polynomial<double> > yy(zero,one);
polynomial<double> x1(0.,1.);
polynomial<polynomial<double> > xx(1,x1);
list<polynomial<polynomial<polynomial<double> > > >
    P(4,ZZero);
P(0) =
    polynomial<polynomial<polynomial<double> > >
    (oneMinusxMinusy,Minus1);
```

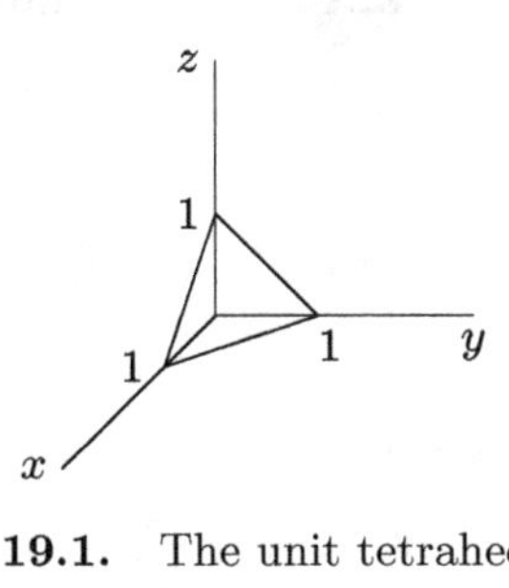

Figure 19.1. The unit tetrahedron T.

```
P(1) =
    polynomial<polynomial<polynomial<double> > >
    (1,xx);
P(2) =
    polynomial<polynomial<polynomial<double> > >
    (1,yy);
P(3) =
    polynomial<polynomial<polynomial<double> > >
    (Zero, One);
```

19.5.2 *Nodal Functions and Their Gradient*

The constant gradients ∇p_0, ∇p_1, ∇p_2, and ∇p_3 are also stored, for further use:

```
point3 gradient[4];
gradient[0] = point3(-1,-1,-1);
gradient[1] = point3(1,0,0);
gradient[2] = point3(0,1,0);
gradient[3] = point3(0,0,1);
```

19.6 Scanning the Finite Elements

19.6.1 *Scanning the Tetrahedra*

We are now ready to scan the tetrahedra in our mesh "m". Each tetrahedron will be denoted by t:

```
for(const mesh<tetrahedron>* runner = &m;
    runner; runner =
    (const mesh<tetrahedron>*)runner->readNext()){
```

In this loop, look at an individual tetrahedron: t.

19.6.2 *Mapping to a General Tetrahedron t*

Look at t: an individual tetrahedron in the mesh (Figure 19.2). How to map the unit tetrahedron T onto t? This is done by an affine

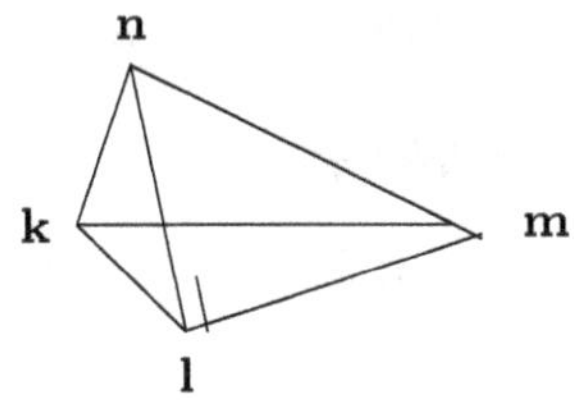

Figure 19.2. A general tetrahedron t, vertexed at **k**, **l**, **m**, and **n**.

mapping:

$$D_t : T \to t.$$

Let's calculate its Jacobian S_t and store it in "S":

```
matrix3 S((*runner)()[1]() - (*runner)()[0](),
          (*runner)()[2]() - (*runner)()[0](),
          (*runner)()[3]() - (*runner)()[0]());
matrix3 Sinverse = inverse(S);
matrix3 SinverseT = transpose(Sinverse);
```

Later on, in the integration, we'll often use $|\det(S_t)|$. Better calculate it once and for all:

```
double detS = fabs(det(S));
```

Furthermore, we'll also need a few more polynomials of three variables (initialized to zero in t):

```
list<polynomial<polynomial<polynomial<double> > > >
    X(3,ZZero);
list<polynomial<polynomial<polynomial<double> > > >
    Y(3,ZZero);
polynomial<polynomial<polynomial<double> > >
    E2(1,Zero);
```

Later on in the loop, when we move on to the next tetrahedron, these polynomials will reset to zero back again, as required. In particular, "E2" will store $\|E\|^2$ in t. This will be useful in the nonlinear term.

19.7 Scanning Blocks and Corners

19.7.1 *Scanning the Blocks*

Recall that our matrix has 3×3 blocks. To scan them, use the new index "iblock"$= 0, 1, 2$:

```
for(int iblock=0;iblock<3;iblock++){
```

19.7.2 *Scanning the Corners*

Inside this loop, use a yet inner loop to scan the corners of t, indexed by "i" (locally in t), or "I" (globally in the entire mesh):

```
for(int i=0; i<4; i++){
  int I = (*runner)()[i].getIndex();
```

19.7.3 *The Electric Field E*

In t, how to approximate the (unknown) electric field E? Here is our current approximation (written explicitly, in terms of its real and imaginary parts, ready for real arithmetic):

$$E_1 \doteq (\text{“X[0]”} + \sqrt{-1}\text{“Y[0]”}) \circ D_t^{-1},$$
$$E_2 \doteq (\text{“X[1]”} + \sqrt{-1}\text{“Y[1]”}) \circ D_t^{-1},$$
$$E_3 \doteq (\text{“X[2]”} + \sqrt{-1}\text{“Y[2]”}) \circ D_t^{-1}$$

(where "$\circ$" means composition). To approximate E in this way, we must sum four polynomials (of three variables): four degrees of freedom: a coefficient from "x" (at the relevant corner of t) times the relevant nodal function (stored in "P"):

```
X(iblock) += x[I+iblock*nodes][0] * P[i];
Y(iblock) += x[I+iblock*nodes][1] * P[i];
```

Once the four corners (and the three blocks) are scanned, E will get well-approximated, and ready for real arithmetic (in t): linearize the real and imaginary parts (which are both differentiable), each on its own.

19.7.4 *Mixed Boundary Conditions*

Moreover, let's define "xBoundary" at the corners of t: 1 at those corners that lie on the left edge of the domain (where mixed boundary condition should be imposed) and 0 elsewhere:

```
          if((*runner)()[i]()[0] <= 1.e-6)
            xBoundary(I) = 1;
        }
```

In this form, "xBoundary" will be used later to assemble the boundary term. This completes the inner loop over the corners of t.

19.7.5 *The Nonlinear Term*

Next, in t, let's calculate the quadratic polynomial

$$\text{“E2”} = \|E\|^2 = |E_1|^2 + |E_2|^2 + |E_3|^2 :$$

```
        E2 += X[iblock] * X[iblock];
        E2 += Y[iblock] * Y[iblock];
      }
```

This completes the loop over "iblock"$= 0, 1, 2$. (This index is now free for further use.) By now, in t, the electric field is well-approximated as

$$E \doteq \text{“X”} + \sqrt{-1}\,\text{“Y”},$$

and ready to get differentiated (in real arithmetic). Likewise, $\|E\|^2$ is up-to-date too: stored in "E2", and ready to be plugged into "f" later.

19.8 Curl and Gradient

19.8.1 *Scanning the Blocks Again*

How to construct the individual rows in A? Recall that A has 3×3 blocks. Let's scan them once again, using the same index: "iblock"$=$ $0, 1, 2$:

```
      for(int iblock=0;iblock<3;iblock++){
```

In this loop, we'll often need the parameter $K_1|\det(S_t)|$. Let's store it, once and for all:

```
double helmNonlin = HELMNonlin * detS;
```

19.8.2 *Scanning the Corners Again*

Inside this loop, let's scan the corners of t again. As before, "i" indexes the corner locally (in t), whereas "I" indexes it globally (in the entire mesh):

```
for(int i=0; i<4; i++){
  int I = (*runner)()[i].getIndex();
```

This way, we're now at the "I"th node in "m", at the nodal basis function

$$\psi_I = p_i \circ D_t^{-1}.$$

In t, this is a composition: D_t^{-1} maps from t back onto the unit tetrahedron, where p_i is a nodal function. Thanks to this, in t, ψ_I is a nodal basis function: 1 at this corner, and 0 at the other three.

19.8.3 *The Gradient*

Thanks to the chain rule, ψ_I has a constant gradient in t:

$$\nabla\psi_I = \nabla\left(p_i \circ D_t^{-1}\right) = S_t^{-t}\nabla p_i \circ D_t^{-1}.$$

Let's store this:

```
point3 gradi = SinverseT * gradient[i];
```

19.8.4 *The Curl*

The electric field E contains three scalar components. Let's focus on E_1 (indexed by "iblock"$= 0$). To help approximate it (in t), design a new vector function (by adding two dummy components):

$$\psi_I^{(1)} = \begin{pmatrix} \psi_I \\ 0 \\ 0 \end{pmatrix}.$$

Its curl is

$$\nabla \times \psi_I^{(1)} = \nabla \times \begin{pmatrix} \psi_I \\ 0 \\ 0 \end{pmatrix} = \begin{pmatrix} 0 \\ (\psi_I)_z \\ -(\psi_I)_y \end{pmatrix}.$$

In t, this is constant. Store it:

```
point3 curli = iblock == 0 ?
  point3(0.,gradi[2],-gradi[1])
:
  iblock == 1 ?
    point3(-gradi[2],0.,gradi[0])
  :
    point3(gradi[1],-gradi[0],0.);
```

In a dielectric material, the curl is not quite constant: "curli" should be multiplied by μ^{-1} (the inverse permeability matrix). But here, stick to vacuum, where μ is just the identity matrix.

19.9 Assembling the Residual

19.9.1 *The Nonlinear Term*

To start, how to assemble the contribution from the nonlinear term to "f"? Integrate $K_1 \|E\|^2 E\psi_I$ in t. Better yet, integrate in the unit tetrahedron T:

```
f(I+iblock*nodes) += helmNonlin *
    point(integral(E2 * X[iblock] * P[i]),
    integral(E2 * Y[iblock] * P[i]));
```

[since "helmNonlin"$= K_1|\det(S_t)|$, where S_t is the Jacobian of $D_t : T \to t$].

19.10 The Matrix Element

19.10.1 *Inner Loop on the Corners*

How does t contribute to the ("I", "J")th element at the present block in A? To assemble this, we need an inner loop over the corners of t:

```
for(int j=0; j<4; j++){
  int J = (*runner)()[j].getIndex();
  polynomial<polynomial<polynomial<double> > >
      Pij = P[i] * P[j];
```

Better yet, calculate the "Pij"s already in Section 19.5.2, and store them for further use (in a 4×4 array of polynomials).

19.10.2 *The Linear Mass Term*

Let's start with the linear mass term. Its coefficient is

```
double helm = HELM * detS / 120.;
if(i==j)
  helm *= 2.;
```

(where "HELM"$= K$). The coefficient $1/120$ follows from the integral over the tetrahedron:

$$\iiint_t (p_i p_j) \circ E_t^{-1} dxdydz = |\det(S_t)| \iiint_T p_i p_j dxdydz$$
$$= \begin{cases} \frac{|\det(S_t)|}{60} & \text{if} \quad i = j \\ \frac{|\det(S_t)|}{120} & \text{if} \quad i \neq j. \end{cases}$$

19.10.3 *Inner Loop on the Blocks*

In a dielectric material, t contributes to all 3×3 blocks. To assemble this, we must loop over not only "iblock"$= 0, 1, 2$ but also "jblock"$=$ $0, 1, 2$:

```
for(int jblock=0; jblock<3; jblock++){
```

But here, in vacuum, the linear mass term contributes to the diagonal blocks only: the (1,1)th, (2,2)th, and (3,3)th blocks only:

```
double helmDiag = iblock == jblock ?
  helm
:
  0.;
```

We are now ready to assemble the linear mass term. Let's leave this for now.

19.10.4 *The Stiffness Term*

And how to calculate the stiffness term? For this, we need the curl and the gradient of ψ_J too:

```
point3 gradj = SinverseT * gradient[j];
point3 curlj = jblock == 0 ?
  point3(0.,gradj[2],-gradj[1])
:
  jblock == 1 ?
    point3(-gradj[2],0.,gradj[0])
  :
    point3(gradj[1],-gradj[0],0.);
```

This way, we already have two curls in t:

$$\text{“curli”} = \nabla \times \psi_I$$
$$\text{“curlj”} = \nabla \times \psi_J.$$

19.11 Assemble the Linear Terms

19.11.1 *Assemble the Curl and the Linear Mass Term*

In t, these curls are constant vectors. Now, do three things:

- Calculate their inner product (in 3-D).
- Add the linear mass term (stored in “helmDiag” above).
- Integrate in t (or, better yet, in the unit tetrahedron T, which has volume 1/6):

```
T addIJ = helmDiag +
    detS / 6. * (curlj * curli);
```

The linear contributions are now stored in “addIJ”. (Later on, “T” will be specified as “matrix2”, so “addIJ” will actually be a multiple of the 2×2 identity matrix.) This is now added to the (“I”, “J”)th element at the present block in A (using the "assemble" function, to be defined later):

```
assemble(I+iblock*nodes, J+jblock*nodes, addIJ);
```

This is the linear part of A. (Don't forget: calculate this once and for all, before the Newton iteration even starts.)

19.11.2 *Assemble the Residual*

Moreover, "addIJ" contributes to "f" as well:

```
f(I+iblock*nodes) +=
    addIJ * x[J+jblock*nodes];
```

This assembles the linear stiffness-mass terms, as required.

19.12 Linearization

19.12.1 *The Linearized Term*

To linearize the nonlinear mass term, differentiate it with respect to c_J (the coefficient of ψ_J). This changes from block to block. For instance, at the (1,1)th block (indexed by "iblock"="jblock"$= 0$), we need the partial derivative

$$\frac{\partial\left(\|E\|^2 E_1\right)}{\partial c_J^{(1)}}.$$

19.12.2 *Differentiation in Real Arithmetic*

But, there is a problem: in complex arithmetic, $\|E\|^2$ is nondifferentiable. Better use real arithmetic; write the real and imaginary parts explicitly:

$$c_J^{(1)} \equiv \left(\Re c_J^{(1)}, \Im c_J^{(1)}\right)^t \quad \text{and} \quad E_1 \equiv (\Re E_1, \Im E_1)^t .$$

In this form, we have a 2×2 Jacobian:

$$\begin{aligned}
&\frac{\partial\left(\|E\|^2\Re E_1, \|E\|^2\Im E_1\right)}{\partial\left(\Re c_J^{(1)}, \Im c_J^{(1)}\right)} \\
&= \begin{pmatrix}\Re E_1\\ \Im E_1\end{pmatrix}\left(\frac{\partial\left(\|E\|^2\right)}{\partial\Re c_J^{(1)}}, \frac{\partial\left(\|E\|^2\right)}{\partial\Im c_J^{(1)}}\right) + \|E\|^2\frac{\partial\left(\Re E_1, \Im E_1\right)}{\partial\left(\Re c_J^{(1)}, \Im c_J^{(1)}\right)} \\
&= E_1\left(\frac{\partial\left(\|E\|^2\right)}{\partial\Re E_1}, \frac{\partial\left(\|E\|^2\right)}{\partial\Im E_1}\right)\frac{\partial\left(\Re E_1, \Im E_1\right)}{\partial\left(\Re c_J^{(1)}, \Im c_J^{(1)}\right)} \\
&\quad + \|E\|^2\frac{\partial\left(\Re E_1, \Im E_1\right)}{\partial\left(\Re c_J^{(1)}, \Im c_J^{(1)}\right)} \\
&= E_1\left(2\Re E_1, 2\Im E_1\right)\begin{pmatrix}\psi_J & 0\\ 0 & \psi_J\end{pmatrix} + \|E\|^2\begin{pmatrix}\psi_J & 0\\ 0 & \psi_J\end{pmatrix} \\
&= \left(2E_1E_1^t + \|E\|^2\begin{pmatrix}1 & 0\\ 0 & 1\end{pmatrix}\right)p_j \circ D_t^{-1}.
\end{aligned}$$

To assemble into the (1,1)th block, multiply this by $\psi_I = p_i \circ D_t^{-1}$, and integrate in t (or, better yet, in T). Let's start with the latter term (that contains $\|E\|^2$). Assemble in not only the (1,1)th but also the (2,2)th and (3,3)th blocks:

```
if(iblock==jblock){
  point J1 = point(integral(E2 * Pij),0.);
  point J2 = point(0.,J1[0]);
  matrix2 Jacobian = matrix2(J1,J2);
  assemble(I+iblock*nodes,J+jblock*nodes,
      helmNonlin * Jacobian);
}
```

The former term, on the other hand, is trickier: it exists in the off-diagonal blocks too. For instance, in the (1,2)th block,

$$
\begin{aligned}
&\frac{\partial\left(\|E\|^2\Re E_1, \|E\|^2\Im E_1\right)}{\partial\left(\Re c_J^{(2)}, \Im c_J^{(2)}\right)}\\
&= \begin{pmatrix}\Re E_1\\ \Im E_1\end{pmatrix}\left(\frac{\partial\left(\|E\|^2\right)}{\partial \Re c_J^{(2)}}, \frac{\partial\left(\|E\|^2\right)}{\partial \Im c_J^{(2)}}\right) + \|E\|^2\frac{\partial\left(\Re E_1, \Im E_1\right)}{\partial\left(\Re c_J^{(2)}, \Im c_J^{(2)}\right)}\\
&= E_1\left(\frac{\partial\left(\|E\|^2\right)}{\partial \Re E_2}, \frac{\partial\left(\|E\|^2\right)}{\partial \Im E_2}\right)\frac{\partial\left(\Re E_2, \Im E_2\right)}{\partial\left(\Re c_J^{(2)}, \Im c_J^{(2)}\right)} + (0)\\
&= E_1\left(2\Re E_2, 2\Im E_2\right)\begin{pmatrix}\psi_J & 0\\ 0 & \psi_J\end{pmatrix}\\
&= 2E_1E_2^t p_j \circ D_t^{-1}.
\end{aligned}
$$

To assemble this into the ("I", "J")th element in the $(1,2)$th block, multiply by $\psi_I = p_i \circ D_t^{-1}$, and integrate in t (or, better yet, in T). Likewise, assemble in not only the (1,2)th but also the ("iblock", "jblock")th block:

```
        point J1(integral(X[iblock] * X[jblock]
            * Pij),
            integral(X[iblock] * Y[jblock] * Pij));
        point J2(J1[1], integral(Y[iblock]
            * Y[jblock] * Pij));
        matrix2 Jacobian(J1,J2);
        Jacobian *= 2.;
        assemble(I+iblock*nodes,J+jblock*nodes,
            helmNonlin * Jacobian);
    }
```

Better yet, products like "X[iblock]" times "X[jblock]" (or "Y[jblock]") could actually be calculated (and stored once and for all) already in Section 19.7.5 (just like "E2", in t). This closes the inner loop over "jblock"$= 0, 1, 2$. Instead, we'll soon open a yet inner loop over the corners.

19.13 Assemble the Boundary Term

19.13.1 *Mixed Boundary Conditions*

On the left edge of the domain, impose mixed boundary condition:

$$\mathbf{n} \times B = E \quad \text{plus a given free vector function}$$

(where B is the unknown magnetic field, and $\mathbf{n}$ is the outer normal vector, pointing from the left edge leftward). On the right-hand side, this will yield two terms. The former will be thrown onto the left-hand side, and contribute to A. On its way, it will pick a minus sign. The latter, on the other hand, will remain on the right-hand side, and contribute to $\tilde{f}$.

How to assemble these terms? For this, detect a boundary triangle, vertexed at three boundary points (indexed globally by "I", "J", and "K") on the left edge of the domain.

19.13.2 *Yet Inner Loop on the Corners*

For this purpose, open a yet inner loop over the corners of t:

```
for(int k=0; k<4; k++){
  int K = (*runner)()[k].getIndex();
  if((i!=j)&&(j!=k)&&(k!=i)
      &&(xBoundary[I] *
      xBoundary[J] * xBoundary[K]==1)){
```

This is a boundary triangle, vertexed (locally) at "i"≠"j"≠"k". Its area is half the norm of the vector product of its edges:

```
point3 jMinusi = (*runner)()[j]()
    - (*runner)()[i]();
point3 kMinusi = (*runner)()[k]()
    - (*runner)()[i]();
double radiation = sqrt(fabs(HELM))
    * l2norm(jMinusi & kMinusi) / 24.;
```

(where "HELM"= K and "&" stands for "×": vector product in 3-D). The factor 1/24 comes from the integral over the unit triangle:

$$\int_0^1 \int_0^{1-x} p_i p_j dy dx = \begin{cases} 1/12 & \text{if} \quad i = j \\ 1/24 & \text{if} \quad i \neq j. \end{cases}$$

19.13.3 *Contribute to Matrix Elements*

In real arithmetic, a complex number is represented by a little 2×2 matrix: the real part on the main diagonal and the imaginary part off-diagonal. Now, what about our mixed boundary term? It is imaginary: since $K < 0$, it has coefficient $\sqrt{-1}\sqrt{|K|}$. For this reason, it should be off-diagonal. Besides, it should also pick a minus sign (Section 19.13.1):

```
T radiationMatrix(
    point(0.,radiation),
    point(-radiation,0.));
```

This should contribute to the ("I", "I")th and ("I", "J")th elements in the (1,1)th, (2,2)th, and (3,3)th blocks:

```
assemble(I+iblock*nodes, I+iblock*nodes,
    radiationMatrix);
assemble(I+iblock*nodes, J+iblock*nodes,
    radiationMatrix);
```

Thanks to "j"↔"k", the ("I", "I")th element gains twice. This is as required: after all, it has coefficient 1/12, not 1/24 (Section 19.13.2).

19.13.4 *Contribute to the Residual*

Besides, the same boundary term should contribute to "f" too:

```
f(I+iblock*nodes) +=
    radiationMatrix
    * x[I+iblock*nodes];
f(I+iblock*nodes) +=
    radiationMatrix
    * x[J+iblock*nodes];
```

19.13.5 *Assemble the Free Function*

On its right-hand side, the boundary condition contains a free vector function too (Section 19.13.1). For simplicity, assume that this is constant: $(1, 1, 1)^t$ on the left edge of the domain. In $\tilde{f}$, this should be integrated over the above boundary triangle. In the residual "f"= A("x") $- \tilde{f}$, this should also pick a minus sign:

```
                f(I+iblock*nodes) -=
                    point(0.,2. * radiation);
              }
```

Why? Well, this is imaginary: in real arithmetic, it should be placed at the second component. And why should it take factor 2? Well, we already divided by 24. Now, thanks to "j"$\leftrightarrow$"k", there is an (implicit) factor 2 too. Together, we now have

$$2 \cdot 2 \cdot \frac{1}{24} = \frac{1}{6},$$

as required in the integral over the unit triangle:

$$\int_0^1 \int_0^{1-x} p_i dydx = \frac{1}{6}.$$

It is now time to close:

```
            }
```

This closes the innermost loop over "k"$= 0, 1, 2, 3$. This assembles the boundary term, as required. The stiffness-mass matrix A is now complete too. It is now time to close the loops over "j", "i", "iblock", and "runner":

```
          }
        }
      }
    }
  } // linearize and assemble Maxwell system
```

19.13.6 *The "Assemble" Function*

In the above, we often called the "assemble" function to add a contribution to the ("I", "J")th matrix element. Like the above constructor, this function should also be written in the sparse-matrix class in Ref. [24]:

```
void assemble(int I, int J, const T& contribution){
```

There are no blocks here: we want to add to the "I"th row, at the "J"th column in our matrix. Now, there are two possibilities. If the "I"th row already exists, then the new contribution should be placed in a new little row and added:

```
    if(this->item[I]){
      row<T> r(contribution,J);
      *this->item[I] += r;
    }
```

If, on the other hand, the "I"th row doesn't exist as yet, then it must be allocated dynamic memory, and constructed from scratch:

```
    else
      this->item[I] = new row<T>(contribution,J);
} // assemble to the (I,J)th matrix element
```

References

1. Abu-Mostafa, Y.S. *Lecture Notes in Machine Learning.* Cornell University, NY, 2011.
2. Averbuch, A., Neittaanmaki, P., Zheludev, V., Hauser, J., and Salhov, M. Image implanting using directional wavelet packets originating from polynomial splines. *Signal Processing: Image Communication,* 97 (2021), 116334.
3. Averbuch, A. Neittaanmaki, P., and Zheludev, V. Directional wavelet packets originating from polynomial splines. *Advances in Computational Mathematics,* 49 (2023).
4. Bransden, B.H. and Joachain, C.J. *Quantum Mechanics.* Prentice-Hall, New Jersey, 2000.
5. Carlson, B. *Lecture Notes on Quantum Mechanics.* 2019.
6. Carroll, S. *Something Deeply Hidden: Quantum Worlds and the Emergence of Spacetime.* Oneworld Publications, London, (2019).
7. Courant, R. and John, F. *Introduction to Calculus and Analysis* (Vols. 1 and 2). Springer, New York, 1998–1999.
8. Einstein, A. *Relativity: The Special and the General Theory.* Martino Fine Books, Connecticut, 2010.
9. Griffiths, D.J. and Schroeter, D.F. *Introduction to Quantum Mechanics* (3rd edn.). Cambridge University Press, Cambridge, 2018.
10. Hrabovsky, C. *Classical Mechanics: The Theoretical Minimum.* 2014.
11. Jones, H.F. *Groups, Representations and Physics* (2nd edn.). CRC Press, New York, 1998.
12. Kadanoff, L.P. *Quantum Statistical Mechanics.* Taylor and Francis, Milton Park, 2019.

13. Karatzas, I. and Shreve, S.E. *Brownian Motion and Stochastic Calculus* (2nd edn.). Springer, New York, 1991.
14. Kibble, T. *Classical Mechanics* (5th edn.). Imperial College Press, London, 2004.
15. Maloney, A. *Lecture Notes on Quantum Theory.* McGill University, Montreal, 2012.
16. Miller, G.L. Riemann's hypothesis and tests for primality. *Journal of Computer and System Sciences,* 13 (1976), 300–317.
17. Rabin, M.O. Probabilistic algorithm for testing primality. *Journal of Number Theory,* 12 (1980), 128–138.
18. Rivest, R., Shamir, A., and Adleman, L. A method for obtaining digital signatures and public-key cryptosystems. *Communications ACM,* 21 (1978), 120–126.
19. Robinett, R. *Quantum Mechanics: Classical Results, Modern Systems, and Visualized Examples* (2nd edn.). Oxford University Press, 2006.
20. Rostamian, R. *Programming Projects in C for Students of Engineering, Science, and Mathematics.* SIAM, Philadelphia, PA, 2014.
21. Schieve, W.C. and Horwitz, L.P. *Quantum Statistical Mechanics.* Cambridge University Press, Cambridge, 2009.
22. Schuller, M.F. *Lectures on the Geometrical Anatomy of Theoretical Physics and Quantum Theory.* Lecture Notes, Independently Published, N.A., 2019.
23. Schwarz, A. *Quantum Mechanics and Quantum Field Theory from Algebraic and Geometric Viewpoints.* Springer, New York, 2024.
24. Shapira, Y. *Solving PDEs in C++: Numerical Methods in a Unified Object-Oriented Approach* (2nd edn.). SIAM, Philadelphia, PA, 2012.
25. Shapira, Y. *Linear Algebra and Group Theory for Physicists and Engineers* (2nd edn.). Birkhouser, Springer Nature, New York, 2023.
26. Shapira, Y. *Classical and Quantum Mechanics with Lie Algebras.* World Scientific, Singapore (2021).
27. Shapira, Y. *Set Theory and Its Applications in Physics and Computing.* World Scientific, Singapore (2022).
28. Sidi, A. *Vector Extrapolation.* SIAM, Philadelphia, PA, 2017.
29. Strang, G. *Introduction to Linear Algebra* (3rd edn.). SIAM, Philadelphia, 2003.
30. Strogatz, S.H. *Nonlinear Dynamics and Chaos: With Applications to Physics, Biology, Chemistry, and Engineering* (2nd edn.). Westview Press (2014).
31. Susskind, L. and Friedman, A. *Quantum Mechanics: The Theoretical Minimum.* Penguin, London, 2015.

32. Susskind, L. *Lecture Notes on Quantum Field Theory and Particle Physics.* Stanford University, California, 2015.
33. Taylor, J.R. *Classical Mechanics.* University Science Books, Herndon, VA, 2005.
34. Vazirani, U.V. *Lecture Notes on Quantum Mechanics and Quantum Computation.* Berkeley University, Los Angeles, 2011.

Index

E

F

W

Z

www.ingramcontent.com/pod-product-compliance
Lightning Source LLC
Chambersburg PA
CBHW050633190326
41458CB00008B/2256
9789819824588